Risk Analysis for Offshore Structures and Equipment

Based on the results of a study entitled "The safety of offshore oil structures and equipment: a new technological and methodological approach" undertaken by ASTEO under contract to the Commission of the European Communities, Directorate-General for Energy (Oil and Natural Gas Directorate). Translation into English by courtesy of Britoil plc, London.

This study was carried out under the direction of ASTEO's Scientific and Technical Committee chaired by Professor Michel Vigneaux, and under the overall supervision of Mr. Jean-Michel Planeix, by the following:

— *Nuclear section*:
Mr. A. Brin

— *Aeronautical and aerospace section*:
Messrs. D. Compard
J.L. Deschanel
P. Lavedrine

— *Petroleum section*:
• Drilling and Production:
Messrs. J.P. Chaix
R. Leroc
A. Leroy
M. Ningler
M. Rogez
J.C. Vaché
• Dispatch:
Messrs. J.P. Auber
E. Bonjour
A. Kermabon
J. Pigeyre
Mrs. M. Planeix-Dumas
Mr. M. P. Scemama
• Inspection Technology:
Messrs. Y. Durand
P. Lepert
J.M. Planeix

— *Liaison*:
Mr. J.H. Guillon

THE COMMISSION OF THE EUROPEAN COMMUNITIES

Risk Analysis for Offshore Structures and Equipment

ASTEO

(Association Scientifique et Technique pour l'Exploitation des Océans)
Immeuble Ile de France,
Paris La Défense

Graham & Trotman

A member of the Kluwer Academic Publishers Group
LONDON/DORDRECHT/BOSTON

First published in 1987 by

Graham & Trotman Ltd.,
Sterling House
66 Wilton Road
London SW1V 1DE
UK

Graham & Trotman Inc.
Kluwer Academic Publishers Group
101 Philip Drive
Assinippi Park
Norwell, MA 02061
USA

for the Commission of the European Communities, Directorate-General Telecommunications, Information Industries and Innovation

EUR 9121 EN

Softcover reprint of the hardcover 1st edition 1987

British Library Cataloguing in Publication Data

Risk analysis for offshore structures and equipment
1. Offshore oil industry —— Safety measures 2. Offshore gas industry —— Safety measures 3. Risk
I. Association scientifique et technique pour l'exploitation des océans
363.1'1962798 TN871.3

LCCCN 87-11892

ISBN-13: 978-94-010-7955-6 e-ISBN-13: 978-94-009-3245-6
DOI: 10.1007/978-94-009-3245-6

Contents

PART II

The application of risk analysis methods to new offshore oil structures and operations

The aerospace industry — Modern procedures for analysing structures have been successfully used for some considerable time and, as in the nuclear industry, have enabled a high level of safety to be achieved.

Drilling support structures — Modern procedures are becoming increasingly widely used.

Drilling — Procedures have in particular been used in the French "Forage 3000" project, and are also being used for drilling rig automation projects.

Production — Risk analysis methods are used for the layout of the various installations and in the case of production by remote submersible stations (e.g. Geondin — North Cormorant — Skuld).

Living quarters — Efforts must be made to integrate the protection and evacuation of personnel in an overall safety study. Proposals need to be made for specific action in this connection.

New concepts — (TLP's, guyed towers, multiple articulation towers). Since experience in the use of new concepts is virtually non-existent, risk analysis is extremely important in cases such as these.

For ease of consultation, the conclusions and recommendations of the report are given immediately following.

CONCLUSIONS

The essential purpose here is to make recommendations in the light of the considerations contained in this study, with a view to defining the safety priorities to be applied when considering projects submitted to the European Community under Regulation 3056.

Without wishing to indulge in a play on words, we should first point out that reliability of supply for member states must, of necessity, be achieved through safety — in the strict sense of the word — of personnel, installations and the environment in the field of offshore oil exploration and production. Clearly the extreme sensitivity of public opinion, the particular attention paid by the media to spectacular catastrophes, particularly where of industrial origin, *can result in stoppage of offshore oil activities* in the event of grave incidents, leading to major production losses. Safety precautions introduced in the immediate aftermath of an accident are generally Draconian, sometimes excessive, and in most cases unsuitable or inefficient. The public authorities, manufacturers and oil exploration and development companies have thus quite rightly taken the lead, and for several years have been developing and using modern methods of risk analysis leading to substantial improvements in safety.

Under Community Regulation 3056, all new aspects of offshore oil-related technology have in practical terms been the subject of projects which have naturally taken into account the safety and reliability of systems, components and items of equipment. This logical procedure for the initial phase of development must be followed by action which is both more broad and more specific, to ensure safety for men, equipment and the environment.

In these conclusions, we present a series of recommendations which aim, from a safety viewpoint, to achieve acceptance or encouragement of projects associated with offshore oil activities capable of meeting criteria of the following type:

— *Phenomenological (fundamental problems)*
— *Methodological*
— *Social*
— *Environmental*
— *Technological*

and not only technological criteria, as has been the case in calls for bid issued since 1974. For each of these criteria, we suggest a number of action proposals for possible adoption.

1. PHENOMENOLOGICAL

This heading essentially covers studies on materials of construction from the viewpoint of fatigue and ageing in offshore environments, particularly in the Arctic, with respect to:

— steel;
— concrete;
— composite materials etc.

and studies of environment conditions.

Taking welded metal nodes as an example, an initial phase of the study was aimed at improving awareness of the fatigue strength of assemblies subject to potential stresses from external factors (waves, winds, currents etc.). These investigations were undertaken on a large scale under a European programme jointly financed by the ECSC.

The second phase, using the theoretical and experimental knowledge acquired in the initial phase of the programme, was designed to provide designers, fabricators and users of offshore welded metal platforms with recommendations of the fatigue design of these structures and precautions required to increase their service life under acceptable economic conditions. Given the participants in the study, the project was to be carried out under the charge of DG XVII, acting as the central liaison body for a major campaign in offshore oil operations. The split of responsibility between the two organizations within the Commission has in fact had an adverse effect on the cohesiveness of the work produced.

The motives of an undertaking such as this may be summarized as follows:

- The reliability of offshore metal structures, and in particular of fixed platforms, which are extremely difficult to repair, is largely governed by the fatigue strength of their welded joints, exposed to a system of alternating stresses resulting from the action of waves, winds and currents.

- Any improvement in the understanding of this field involves mutually supportive experimental and theoretical investigations leading finally to an engineering design tool providing forecasts of service life based on the physical reality of the phenomena.

These investigations are of interest to all industries working in offshore oil production (oil companies owning the installations, engineering and construction companies, insurance companies) and, of course, to organizations responsible for regulatory and supervisory activities. Hence the centralizing body of these activities must be DG XVII.

The results expected from this type of research must lead to the production of a processing tool, providing the owners of offshore installations with the means, at any time, of evaluating the quality and reliability of their installations in the face of safety requirements and regulations, to enable the designers of offshore metal structures to produce safer and less costly designs.

2. METHODOLOGICAL

Essentially, the development of modern methods of risk analysis and procedures must be encouraged to make them better known to operators, and to improve studies on safety and means of inspection and maintenance. Co-ordination is required in this field, to improve the presentation of existing means of calculating risks, the development of reliability coefficients, and the available means of inspection. Action to be taken in the design of platforms, and more particularly on new concepts, could relate to the following points:

A. Development of a methodology to take into account the behaviour of the structure throughout its life: design, construction, service.

The following aspects would accordingly be incorporated:

— Redundancy;
— Residual structural safety after damage;
— Possibilities for repair and structural safety after repair;
— Inspection parameters: detectable faults, allowable faults, frequency;
— Continuous and semi-continuous inspection methods.

B. Development of a simple means of calculation, a "master programme", making use of the range of structural analysis programmes in existence, and taking into account:

— probabilities of structural safety in the intact state, following damage and following repair
— probabilities for length of service life in the various situations. This development should involve:

- studying and defining, with sufficient precision, the concept of redundacy;
- studying and defining, under the same conditions, the concepts of "detectable faults" and "allowable faults", with regard to the envisaged methods of inspection and the frequency of inspection;
- studying existing codes (offshore structures, bridges, onshore concrete, handling), to take advantage of experience gained in various fields;
- researching values of a statistical nature relating to the mechanical strength (in general sense) of the materials used and the service behaviour of typical structures.

C. Development and refinement for practical use of inspection methods such as:
— vibro-detection;
— acoustic emission;
— pressure detection;

— cumulative fatigue records;
— fatigue indicators.

3. SOCIAL

Action to be undertaken in this field relates to:

- Improvement in working and safety conditions;
- Personnel training.

In both case, the required results will be seen to be achieved by technological progress which will, in addition, bring an improvement in productivity.

Improvement in working and safety conditions
The study examined in particular one of the most important operations in the petroleum sector and paradoxically that which has seen the least development, viz. drilling. A drill man from the 1950s would not feel out of his depth with a 1983 drilling rig.

Progress is urgently required in the following central areas, in all cases with a view to improving safety:

— Work on the drilling floor should be made less unpleasant: it is generally carried out in an aggressive environment with a high noise level, extreme temperatures, vibration and major physical effort. These working conditions generate accidents.

Drilling operations should therefore be automated as far as is possible: this should also enable numbers of personnel living on offshore platforms to be reduced, further militating towards an improvment in safety.

— Drilling parameters should be optimized to improve productivity. This would result from research and development in "MWD" methods (Measurement While Drilling) enabling direct information to be obtained at the surface in real time on the hole bottom conditions (pressure, temperature etc.) enabling action to be taken on the parameters for use of the equipment (see Table I: bottom measurements while drilling).

These techniques will enable drilling to be carried out under totally safe conditions, with particular reference to blow-out risk prevention.

Personnel training: This is a fundamental point in safety, and both operators and States responsible for regulatory activities must do everything necessary in this connection. Technological developments are, however, required and could possibly be supported by DG XVII: these would relate to the investigation and development of simulation methods to provide for personnel training in methods, procedures, and the use of equipment for the prevention of accidents, and corrective action in the event of accidents.

Simulation units are in fact available, reproducing the drilling foreman's position and enabling personnel to be trained in fighting blow-outs. These procedures are very widely used in the aeronautical and nuclear industries, where personnel have to be trained to handle extremely high-cost equipment and where errors are unacceptable. These same conditions apply quite specifically to the offshore oil exploration and production industry. The Community could also be involved in issuing European licences to operators relating to the principal functions in oil exploration and development (drillers, shift managers, wire line operator, drilling rig safety officers, production unit safety officers etc.).

4. ENVIRONMENTAL

As in the case of manpower training, it is up to States and operators to organize pollution prevention and pollution control methods in the event of accidents. DG XVII cannot, however, ignore this subject and therefore, in line with its terms of reference and its specific resources, must participate in the protection of the environment against potential damage resulting from offshore oil exploration or development activities. Although this point is not included in the study, we nevertheless felt that it should be considered in these conclusions, since it could be a consequence of incidents such as blow-outs.

The specific means of action in question relate to support for technological projects possibly concerning the following aspects:

A. *Detection and evaluation of potential hazard*
Remote detection (i.e. by satellite or conventional air methods) should first be improved, and efficient models should be promoted for moving hydrocarbon slicks.

B. *Control methods*
Two processes are used offshore to combat pollution:

- Slick containment with a view to recovery;
- Chemical dispersion.

Containment and recovery can only be achieved under acceptable sea conditions (below force four) and provided the hydrocarbons are not too widely dispersed.

Progress still needs to be made on dispersants, which must not be excessively aggressive on the environment.

In the absence of recovery vessels effectively "gulping" pollution over a broad path in their direction of travel, the relevant principle not as yet having been perfected, the method consists of containing the slick, to allow recovery pumps to be used.

This containment is achieved by tugs or auxiliary vessels setting booms into position. Loading, deploying and placing these heavy and cumbersome items of equipment under windy and difficult current conditions, are tricky operations. The oil then has to be recovered using various types of equipment.

DG XVII could support technological projects in this field of recovery (on oil recovery vessels, booms for use in heavy seas) and in the field of dispersants.

5. TECHNOLOGICAL

We shall not go into this subject in detail, save to comment that on each of the general aspects examined above, technological progress is needed in the main areas of action summarized below:

Priority action by DG XVII in supporting projects on:

— Studies on materials of construction;
— Refinement of design codes to make use of risk analysis;
— Improvement of inspection methods;
— Automation of drilling operations;
— Optimization of drilling parameters (MWD techniques);
— Investigation and development of simulation methods for offshore drilling and production personnel training;
— Development of dispersants for hydrocarbon pollution control offshore;
— Investigation and development of means of containment (booms) and recovery (recovery vessels).

Liaison to be carried out by DG XVII
— Programme harmonization for studies on materials of construction performed by various departments or sections within the Commission.
— Standardization of occupational licences for drilling and production personnel.

Promotional activities to be performed by DG XVII on methods of:

— Application of risk analysis;
— Training of drilling and production personnel;
— Pollution control.

The above list is clearly indicative only, and is not claimed to be exhaustive. We would point out once again that pollution control facilities are not covered by this study: they would, however, merit specific development.

TABLE I. Bottom measurements whilst drilling (conceivable applications by drillers).

		Immediate use				*Deferred use*		
	Safety			*Drilling operations*		*Inspection*	*Technical Documentation for*	
	of persons	of the environment	of equipment	Technical and financial optimization of drilling	Implementation of technical programme	Compliance with technical recommendations. Time analysis. Compliance with regulations	Immediate use on site for analysis of developments in phenomena	Subsequent use at base
Weight				+	+	+	+	+
Torque				+	+	+	+	+
Rotational speed (turbines)				+			+	+
Mud pressure inside drill string	+	+	+	+	+	+		
Mud pressure in annular space	+	+	+	+	+	+	+	+
Mud resistivity	+	+	+				+	+
Temperature	+	+	+				+	+
Ultrasonic gas detection	+	+	+				+	+
Directional well – Angle		+	+	+	+	+	+	+
Directional well – Azimuth		+	+	+	+	+	+	+
Directional well – Toolface		+	+	+	+	+	+	+

TABLE II. Bottom measurements whilst drilling (conceivable applications by geologists).

Type of measurement	*Immediate use*								*Deferred use*			
	Detection in mud		*Identification*		*Detection in formation*				*Evaluation*	*Other*		
	Salt	*Fluid show*	*Lithology*	*Formation fluids*	*Abnormal pressures*	*Loss zones*	*Evapor. series proximity*	*Cracks*	*Res. design (Ø-K)*	*Correlations*	*Geochemistry*	*Drilling Log Program Modif.*
Drilling rate			+		+			+		+		
Gamma Ray (GR)			+	+	+				+	+	+	
Sonic (Δt)			+	+	+	+		+	+	+	+	
Resistivity (R)			+	+	+			+	+	+	+	
Temperature (θ)		+		+	+	+	+	+			+	+
Mud resistivity (Rm)	+	+		+	+							+
Torque			+					+		+		

"Pétrole et Techniques No. 285 – December 1981 AFTP Congress – May 1981.

Part I

The New Procedures of Risk Analysis by the Fault Tree Method Used in the Nuclear, Aeronautical and Aerospace Fields

Chapter 1

Description of the Methods of Risk Analysis Used in the Nuclear Field by the Commissariat A L'Energie Atomique (CEA)*

INTRODUCTION

To date, regulations on safety in the nuclear industry have been based on deterministic rather than probabilistic criteria. The probabilistic approach is, however, used to determine systems reliability.

The determinist principles of design are summarised in Appendix 1, Fig.A.1.1, and the contribution made by the use of probabilistic studies is summarised in Fig.A.1.2.

THE PROBABILISTIC APPROACH

Following a summary review of the components reliability theory, Appendix 1 goes on to detail the methodology adopted by the CEA. Some of the tables are devoted to common modes and human error, the limiting factors in systems reliability. It is in these two fields that data are, in fact, most lacking. In view of the multiple redundancies, however, no accident is the result of a single event or of failure of a single system: there has to be a chain of undesirable events, or an 'accidental sequence'. These sequences are studied by the event tree method presented in Figs.A.1.20 and A.1.21.

Only risk analysis (Probabilistic Risk Assessment (PRA)) can, however, take into account all undesirable events and the consecutive accidental sequences, and study their consequences in order to finally add the probabilities of the branches to the unacceptable consequences. (For a comparison of the two approaches, see Table A.1.5. The first risk analysis was WASH 100 (the now famous Rasmussen report) from which the curves shown in Figs.A.1.22 and A.1.23 became well-known. Since then, other types of risk analysis have been used, particularly in West Germany (Figs.A.1.24 and A.1.25).

CONCLUSIONS

A totally probabilistic approach to reliability, having the advantage of being exhaustive and realistic, would be conceivable, subject to the definition of criteria: this is in progress. However, the parallel performance of risk analyses allows us to skirt this problem of determining the acceptable risk criterion.

---

*For a detailed description, see Appendix 1.

Chapter 2

Systems Reliability and Technological Risks in the Aeronautical and Aerospace Industries

SYSTEMS RELIABILITY

Although Systems Reliability techniques first appeared in Bell Telephone's engineering offices in 1962, their application over the next 15 years was limited to the fields of advanced technology and risks of a major level of gravity. This was for two main reasons:

(1) These military, aerospace and electronuclear activities, in spite of their dissimilarities, exhibited certain critical elements in common: major technical, financial and even human risks. But above all, they shared a lack of experience in their own technologies, hence the absence of statistical data and suitable safety codes. Since reliability could be based neither on regulations nor on insurance, the situation was bound to move towards Analytical Risk Management.

(2) The capital expenditure necessary to develop and perfect efficient tools for investigation and processing was restricted to advanced programmes at national level.

Since 1974, under these conditions, Aérospatiale have been working on the development of a Systems Reliability activity designed to cover the National Strategic Nuclear Force reliability requirements.

Aérospatiale's know-how in this field has been developed primarily within the context of its industrial work on ballistic systems: SSBS's launched from underground silos and MSBS's from submerged nuclear submarines.

In addition, work on transferring these new methods to different sectors of activity undertaken since 1978 has related successively to:

(a) non-nuclear military systems, mostly in marine environments (FOCH aircraft carriers, the Masurca system, the towed sonar, etc.);

(b) unfamiliar fields for the systems approach (Ses-Dumez housing, river pollution problems for the Agence du Bassin d'Artois-Picardie, civil defence, urban defence);

(c) computerised systems in widely varying applications (distribution, computer-aided design, bank management, local authority data processing, etc.).

Experience thus gained has shown that while the methods of risk analysis evolved (e.g. fault trees) were well suited to the study of processes and activities outside the conventional fields of application of Systems Reliability, the main contribution made by this transfer of technology was to be found not in the popularisation of these advanced methods, but rather in making available to technical and financial management a rational procedure providing an aid to decision-making.

This has led to the development of a modular approach, which can be adapted to suit requirements, and comprises three stages.

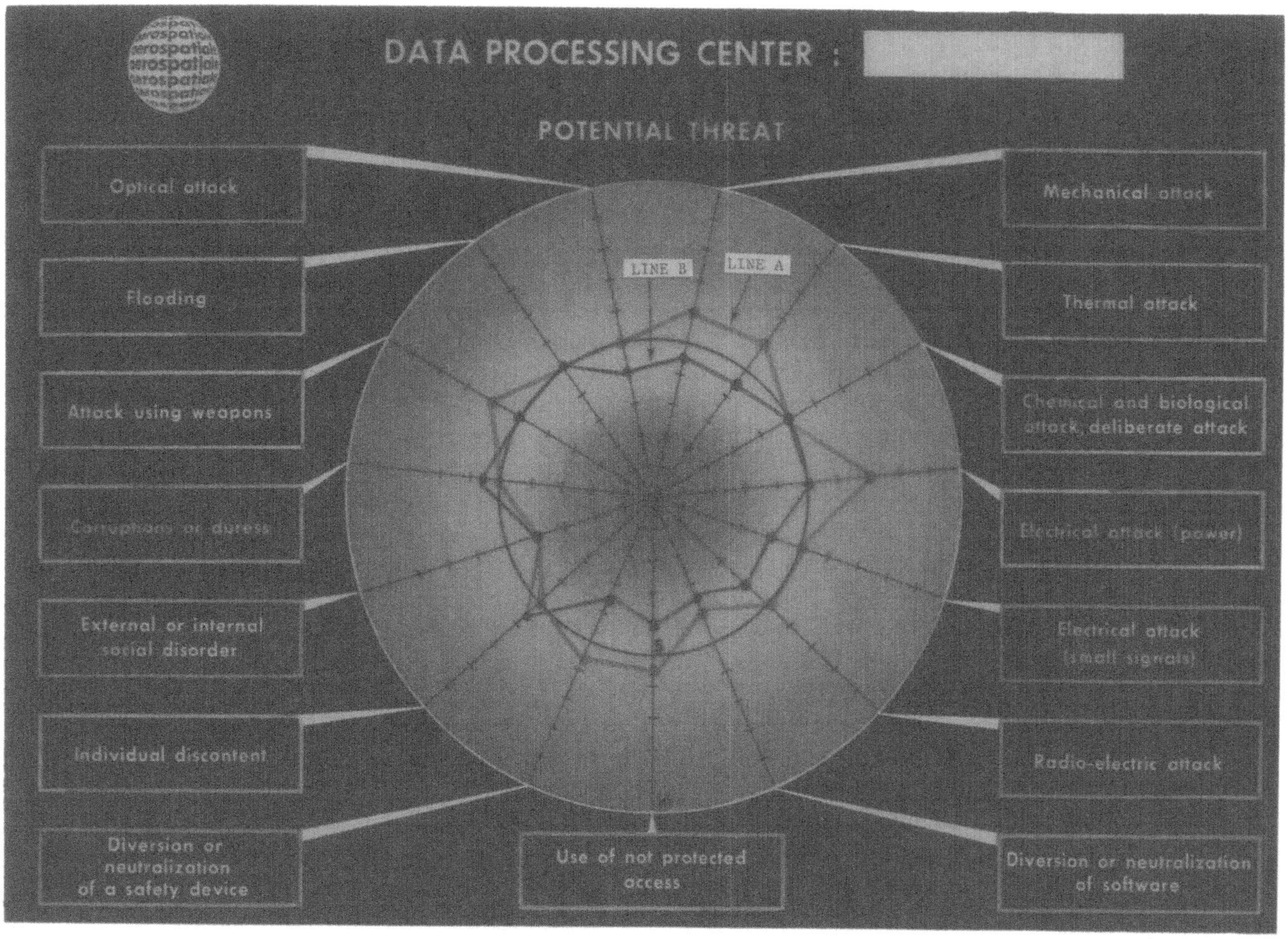

Fig. 1. Data processing centre - potential threats

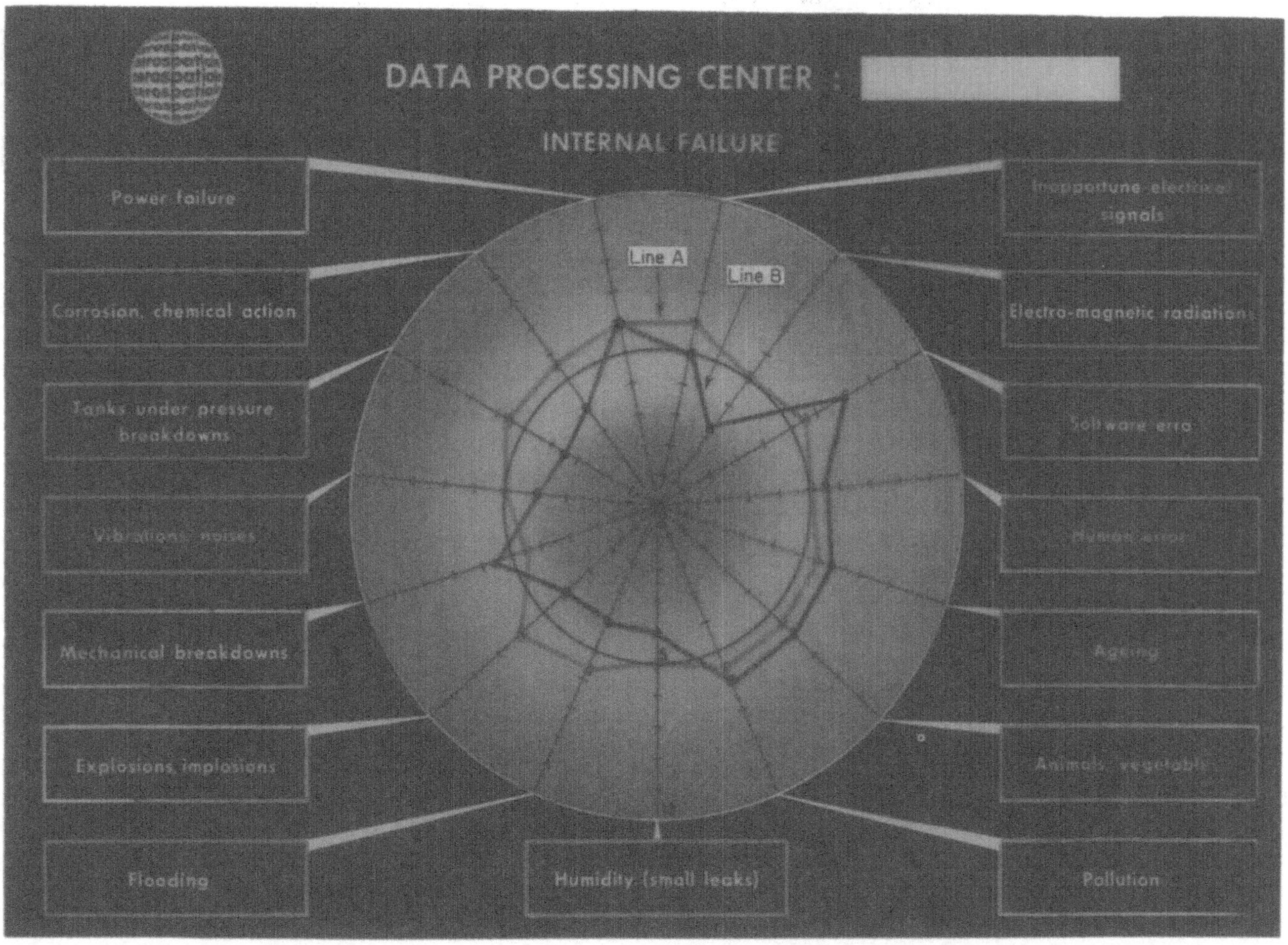

Fig. 2. Data processing centre - internal faults

Table 1 Products suited to the needs of aids to decision-making

Product	**Object**	**Already used in:**	**Conceivable applications in the oil industry**
Site analysis	To identify the system, enabling risk rosettes to be drawn	Data processing centres Urban authorities Automobile industry	Exploration Production
Risk rosettes	To ascertain the situation vis-à-vis risks involved - both accidental and wilful	Urban authorities Data processing centres Manufacturing plants River pollution problems	Exploration Production Despatch
Potential loss evaluation	To ascertain the financial damage suffered in the event of accident	Data processing centres Local authorities	Exploration Production Despatch Distribution
Industrial product safety	To protect the manufacturer against the legal and financial repercussions of accidents involving his products	Pre-fabricated houses	Distribution
Risk analysis	To evaluate the probability of occurrence of possible accidents	Navy Water treatment GDF Electrolyser	Drilling Production Despatch Distribution
Safety strategy	To help define safety policy and its contractual and financial implications	Weaponry systems Water treatment Data processing centres Urban authorities etc.	All functions

The Review

This is designed to provide the manager with a 'photograph' of the system regarding reliability, in the form of risk 'rosettes', presented in Figs.1 and 2 in the computerised context. For different types of accidental or subversive risk, rosettes reveal:

(1) the degree of exposure of the system (line a);

(2) the degree of protection obtained (line b).

The mean of these two values thus provides the level of reliability achieved, with a notation of between 0 and 10.

The 'rosettes' are prepared on the basis of a questionnaire completed by the specialists in the establishment in question. They enable the manager to define priorities for action.

Two 'rosettes' are shown on the following pages: Fig.1 shows the treatment of potential threats and Fig.2 shows the treatment of internal faults.

Diagnosis

This is designed to provide a financial quantification of the risks involved, based on an estimate of the potential direct and indirect losses, extending from material damage and bodily injury in cases of accident to loss of earnings in cases of interruption in service, and taking into account possible damage to the company's image.

This stage enables the manager to define a Risk Management policy and to allocate the safety budget to prevention (technical measures) or protection (insurance) with the aim of improving efficiency. Large insurance companies are currently trying out this approach with a view to improving cover for data processing risks.

Processing

This is designed to identify the best means of improving reliability levels, and uses the various methods of systems reliability (fault trees, Petri networks, analytical graphs, field surveys, etc.) as suitable for evaluating the cost/benefit ratio of the action proposed.

This modular Risk Management strategy leads to the provision of products suited to the modern requirement for aids to decision-making, which have already been in use for some time in extremely wide varying fields of activity, as shown by Table 1.

In most cases, application of the Reliability Strategy produced in this way leads to a Training Strategy, eventually enabling in-house teams of specialists to carry out their own diagnosis of the system or activity being developed or used by them.

TECHNOLOGICAL RISKS: REQUIREMENTS AND STRATEGIES

Although dealing with technological risk is explicitly acknowledged as an undeniable need with respect to aeronautical structures, it raises a number of widely varying problems:

(a) depending on the type of risk involved ('major' risks being liable to lead to catastrophic, even irreversible damage, or 'minor' risks when considered from the national economy or policy standpoint, but nevertheless of a formidable nature for individuals or commercial organisations).

(b) depending on the viewpoint from which they are observed (i.e. the manufacturer marketing a technical device, the user of such device, himself a manufacturer, or the manager in an administrative or commercial organisation).

Is it possible to establish the basis of a general approach for dealing with risk, i.e. to formulate the 'safety problem' clearly, but in a sufficiently universal fashion to enable each manager quickly to extract the term he requires and formulate the policy he needs to adopt to best meet his requirement?

When expressed in these terms, our enquiries inevitably lead us to consider an existing discipline, i.e. Risk Management. However, neither the objectives nor the means proposed by risk management specialists seem to be sufficiently complete: they are tainted with limitations from the outset. Indeed, risk management developed as a response to a need for advice for manufacturers with respect to insurance. Insurance is not the universal means of dealing with risk, but simply a means of providing financial cover for contractually defined risks. This definition makes it impossible to resolve all safety problems incurred by industrial management, for three main reasons:

(1) Basing its necessary financial balance on the notion of mutuality of risk, insurance covers certain standardised risks by means of an individual contract. It cannot, therefore, be based on a precise identification of the risks incurred by each of its clients.

(2) The current expansion in legal consultancy organisations demonstrates the inadequacy of guarantees provided by insurance for the area of boundaries of liability between the supplier and his client.

(3) Finally, and in our view, most seriously, insurance does not set out to prevent accidents from occurring. To introduce terminology that we shall be using further in this study, it provides not prevention, but protection. Optimising the way of dealing with risks will often lead a manager to make an enlightened choice between two possible policies - that of prevention (reducing the probability of accident) and that of protection (reducing the consequences of an accepted accident).

Following this avalanche of criticism, we must nevertheless pay tribute to insurance: it does provide an absolute guarantee of financial cover for risks defined by contract. And in this respect, it is irreplaceable and its inclusion in the procedures which we shall be proposing is beyond question.

We cannot, however, accept the initial approach of the risk manager, which we find inadequate. The foundations for our reflections on the notion of risk must be sought elsewhere. We shall, in fact, draw them from a recent, but nevertheless, well-established, discipline in the 'major risks' sector (the weaponry, nuclear and aeronautical fields) where prevention is essential and any idea of contractual protection is excluded for reasons of the very extent of any potential accidents. This is the Systems Reliability methodology.

Following an introduction to this methodology, we shall develop general thoughts on the problem of reliability in the use of a technological product. Finally, we shall establish a proposed strategy for dealing with technological risks.

GENERAL INTRODUCTION AND THOUGHTS ON SYSTEMS RELIABILITY

Sources

The 'Systems Reliability' concept was born around 1965 from the encounter between a need and a tool. The need was for systematic analysis of reliability problems finally perceived for certain 'major projects': weaponry systems, the exploration of space, large industrial complexes, nuclear power. These activities, in spite of their differences, shared certain common 'criticalities': the existence of 'major risks' (financial, human, etc.) in each of these specialities, and above all they shared the lack of experience in these new technologies, hence an absence of statistical data and suitable safety codes.

In the absence of experience, there was no alternative but to deal with risks analytically. Added to this requirement was a more general need for a systematic analysis of performance which, a few years previously, had given birth to specialist techniques such as operational research, analogue simulation, etc.

The tool was the fault tree. First tried by Bell in 1962 for the Minuteman missile and taken over by Boeing (again for military systems rather than civil aviation), this analytical technique attained notoriety in the famous Rasmussen report (on the safety of American nuclear power stations). Unfortunately, the ensuing polemic (justified by the frequently questionable assumptions and conclusions of the report) turned on the method itself, which 10 years later, passions having calmed, no-one now regards as very serious. Nevertheless, the publicity given to the fault tree method by the Rasmussen report was almost sufficient to kill it. Where did the fault lie? With the method itself, Rasmussen's use of it, or the virulence of its detractors? The responsibility undoubtedly has to be shared, but in our view, it derives from a fundamental error: that of seeing in the fault tree technique only a <u>quantitative</u> <u>method</u>, its sole objective being to determine the probability of accident.

This quantification was not, however, the only requirement, and nor, therefore, was it the only advantage of the method.

Since the tool has to be adapted to the need, the method had to be incorporated in a coherent methodology at complete system level, taking into account both the environment of the system (the technological, natural and human environment) and its service life (commissioning, maintenance, etc.). This methodology, curiously enough, appeared in the form of a US standard laying down the various safety measures to be adopted throughout the project life, for all new weaponry systems.

This was the 'reliability standard' or 'reliability programme', an explicit formulation of the organisation of methods available. The systems reliability approach is based on:

(a) a definition: 'A system is a combination of hardware, software and human components, organised to interact to fulfil a given task'.

(b) and an assumption: 'There is no such thing as zero risk, so reliability applies if the probability of occurrence of each of the potential accidents is below a pre-determined acceptability level taking into account the gravity of the accident in question'.

Figure 3 illustrates the typical outline of a reliability standard. This outline has a particular fundamental feature: a large number of 'return loops' at all stages. This ensures that the reliability standard will at all times answer to the system in question, the tests undertaken and the specified procedures for use.

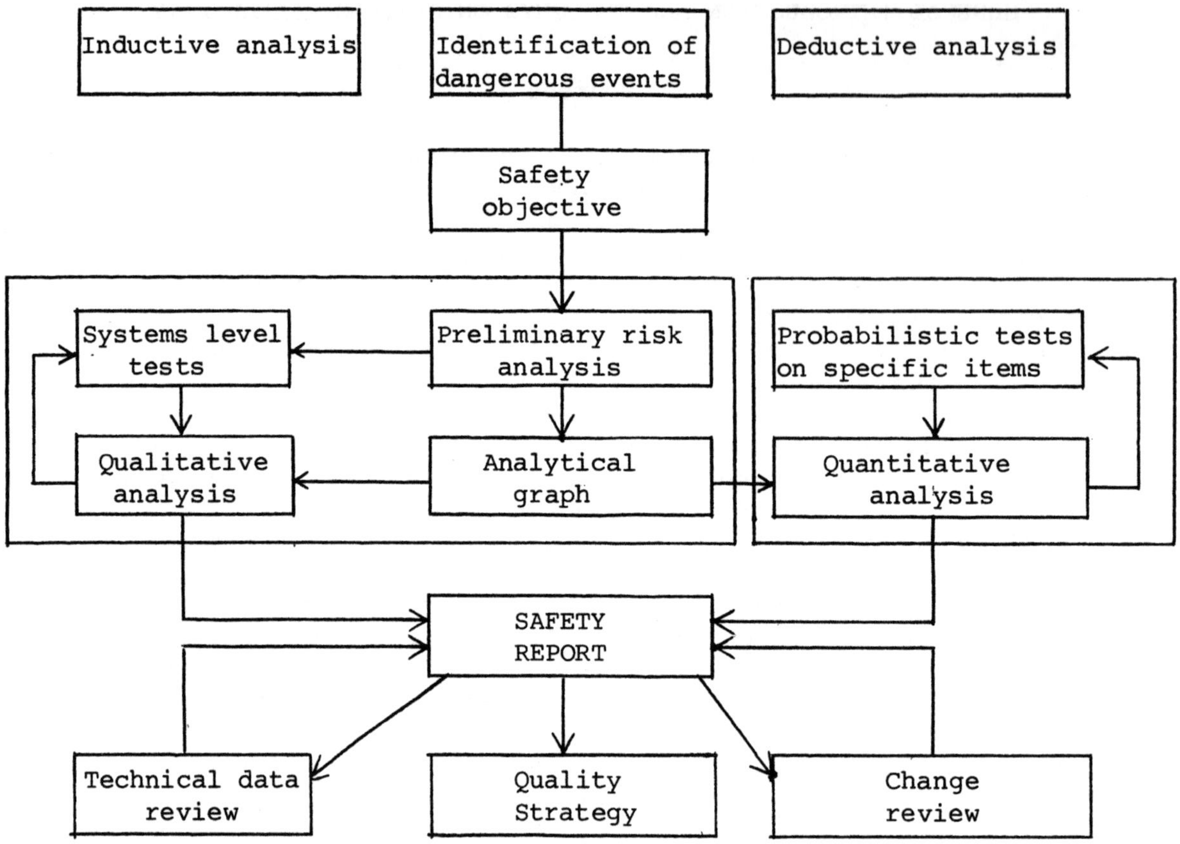

Fig. 3. Typical outline of a reliability standard

Let us now take a look at each stage in this reliability standard.

(1) For the 'reliability' task, the reliability objective is equivalent to the general specifications in the 'functional' task of the system. It defines:

(a) the list of 'dread events' for the system;
(b) the acceptability level for the probability of occurrence of these events;
(c) the systems environment to be taken into account.

A priori selection of the acceptability level (i.e. the probability of occurrence regarded as permissible) of a dread event is a delicate matter. This selection is generally the result of a probability-gravity graph drawn by the analysts, designers and decision-makers, and constituting various classes of 'acceptable' probability as a function of the consequences of a potential accident. The unit chosen to quantify these consequences is not universal. Depending on circumstances, it may be material, financial, human or other loss.

The notion of environment should also be stressed: does reliability still have to be guaranteed under conditions of natural disaster, for example? Is the system in question located in a stormy or oceanic region (humidity), close to a railway line (vibration), etc.?

(2) Preliminary risk analysis (PRA) is an essentially qualitative approach, its object being first to establish a reliability rationale for the system, and secondly to define an initial 'outline' for investigation, providing as full as possible a schedule of conceivable 'accident paths'.

PRA may use a number of methods (failure mode and effect analysis (FMEA), failure mode and effect criticality analysis (FMECA), and analytical graphs) and may be based on various sources of information (analysts' experience, collections of technical data and incidents having occurred previously on similar systems, specialists' appraisals, specific tests). Any tests undertaken at this stage are, of course, qualitative and not probabilistic.

(3) The analytical graph. Depending on the method used, PRA will lead to either a graph or an analytical matrix. Irrespective of the form of document, it provides:

(a) a schedule of 'initiating events' which may be internal to the system (malfunctions), external (aggression), or deliberate if we consider problems of vulnerability (sabotage, intrusion).

(b) conceivable paths (i.e. those which cannot be excluded between these various initiating events and the dread events (reliability objective).

The analytical graph provides a first and important aid to decision-making. Identified accident paths are duly classified into different levels of probability and gravity. Depending on the budget and the time allowed, a decision may be taken either to continue the study on all fronts, or to concentrate efforts on the most critical points.

(4) Qualitative analysis is based on the analytical graph (or matrix) and consists of taking the paths in detail and ensuring that an adequate number of barriers is placed on each of these paths. These barriers are intended to prevent either the occurrence of the initial incident (or aggression), or its aggravation into a dread event. They may be technical devices, software protection, user procedures or quality control.

Qualitative analysis is based on documentary research and collaboration between users, analysts, designers and decision-makers, in collecting and organising the necessary information. It is essentially inductive, i.e. it analyses the incident paths from the initiating events onwards. It may use particular methods: cause and effect graphs and event trees.

(5) Quantitative analysis is intended to provide the probabilistic quantification of the occurrence of dread events. This quantification involves three stages:

(a) the establishment of a mathematical model to represent the paths envisaged. This is the fault tree, in either its simple form or in the more developed form of 'generalised' or 'dynamic' fault trees.

(b) the collection of probabilistic data using reliabilistic data, empirical statistics, specialist calculations and results of probabilistic tests.

(c) the actual processing, including qualitative processing of the tree (to obtain the list of combinations of initiating events or failures which will suffice to cause the dread event) and quantitative processing, providing probabilistic results.

(6) The reliability report. The various studies stated above together form the initial reliability report (the IRR). It is important to note here that the quantitative results are not absolute values for the risks incurred, but instead relate to the increments to these risks. Indeed, to achieve a precise quantification of the risks would involve extremely costly

investigations and tests, without giving any major benefit: it is, however, generally possible to obtain comparable incremental results using penalty assumptions.

Once again, the contribution made by the reliability standard is to provide an aid to objective and rational decision-making. The final decision is always political, taking into account factors other than safety, with particular reference to operational reliability, cost, completion times etc.

The initial reliability report is not an end in itself, for three main reasons.

(a) The reliability study has been produced on a 'documentary model' of the system. Using a suitable quality plan, therefore, the true system must be checked with its documentary model to ensure they are both identical.
(b) The system, once commissioned, will undergo development by means of successive modifications. These modifications should be studied and checked to ensure that they do not have a disastrous effect on the initial reliability level. This is the object of the change review.
(c) The running of the system provides information on the assumptions in the form of minor breakdowns and incidents. These assumptions are monitored during the life of the system by technical data review.

It is therefore desirable, even essential, for the reliability report to review the system throughout its life.

Reliability in Use of a Technological Product

Day-to-day experience reveals three major types of accident situation:

(1) manufacturing faults (e.g. swarf in the stub axle of a car);
(2) misuse (e.g. where lifting tackle breaks due to overloading);
(3) use in an inappropriate environment (e.g. an incandescent lamp used in an explosive atmosphere).

This rudimentary classification reveals two difficulties inherent in dealing with technological risks:

(1) the three situations above are the result of error or ignorance which are clearly not chance and cannot, therefore, be dealt with by reliability studies on a statistical basis;
(2) whilst the first is clearly the responsibility of the manufacturer, the two latter situations raise a delicate problem: can, and must, the manufacturer imagine all abnormal uses which might conceivably be made of a product supplied by him? As for the user, can he be held responsible for errors which he commits due to a lack of information?

As we have seen, the systems reliability approach arose from the problems of complex systems, where an overall contractor was responsible not only for supplying the equipment, but also for user procedures and for the training of user personnel, the latter being subjected to military-type training and discipline. Hence reliability was entirely in the hands of the contractor.

While the task was therefore clearly defined, it was far from simple: analysing the gestation phases of a system, we find one of the first difficulties in dealing with reliability. Indeed, we come across three successive 'models' of the system:

(1) The 'design model': the set of documents, design notes, plans etc. by which the system is defined. It is to this model that risk analysis relates.

(2) The 'as built model': the technological entity produced according to the above model, but for various reasons (limits on physical knowledge, manufacturing tolerances etc. ...) not perfectly identical to this model.

(3) Finally, the 'model perceived by the user', who is not familiar with the details of his system but who, from his general knowledge and his experience of the same type of system, has created his own personal idea of the system.

This breakdown is given in Fig.4, in which the arrows reveal the potential differences between the various models.

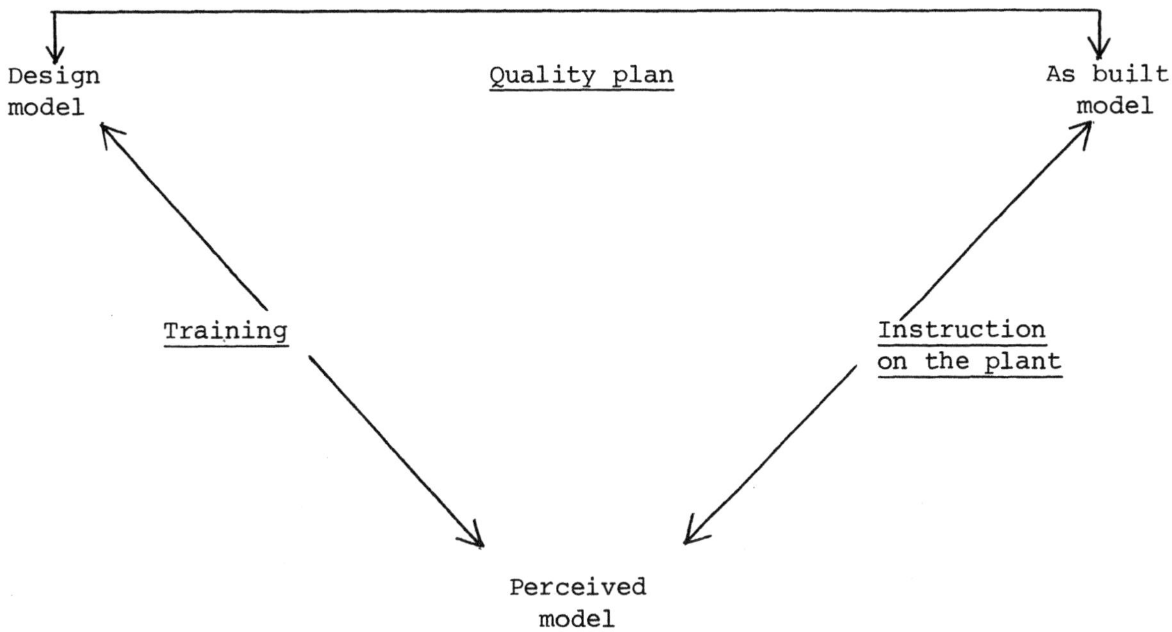

Fig. 4. Potential differences between various models

Differences between the design model and the as built model will lead the latter to give a different performance from that required, giving rise to potential sources of accident. These differences can be dealt with by preparing a careful and intelligent Quality Standard.

Differences between the as built model and the perceived model will lead to poor use of the system. They can be reduced, or controlled, by operator instruction on the as built equipment.

Finally, the perceived model can be brought closer to the design model by personnel training using the documentation supplied by the contractor.

If we look at reliability in the client's use of a technological device manufactured in series by a contractor, the problem becomes more complex. On investigation of the boundaries of responsibility, requiring a more precise analysis of accidental sequences, we have to divide reliability into three categories:

(1) the probability of failure of an electronic component is a function of its service life: this type of failure comes under product reliability;

(2) the probability of breakage of a mechanical device is a function of the number of stresses to which it is subjected: this type of failure comes under reliability in use;

(3) finally, other types of failure may be caused by the environment of the system - its climatic environment (humidity), and the industrial environment (vibration): this type of failure derives from environment-related reliability.

The introduction of these three notions leads us to complete the scheme proposed earlier:

(a) The design model comprises:

- the technical definition of the product;
- its theoretical use (specification of requirement);
- the specified environment.

(b) The as built model comprises:

- the product as built;
- all possible uses of the product as built (generally, the product as built has broader possible uses than those envisaged by the designer).

(c) The living model (the term we shall use henceforth in preference to the 'perceived' model) comprises:

- the perceived product;
- its actual use;
- its actual environment.

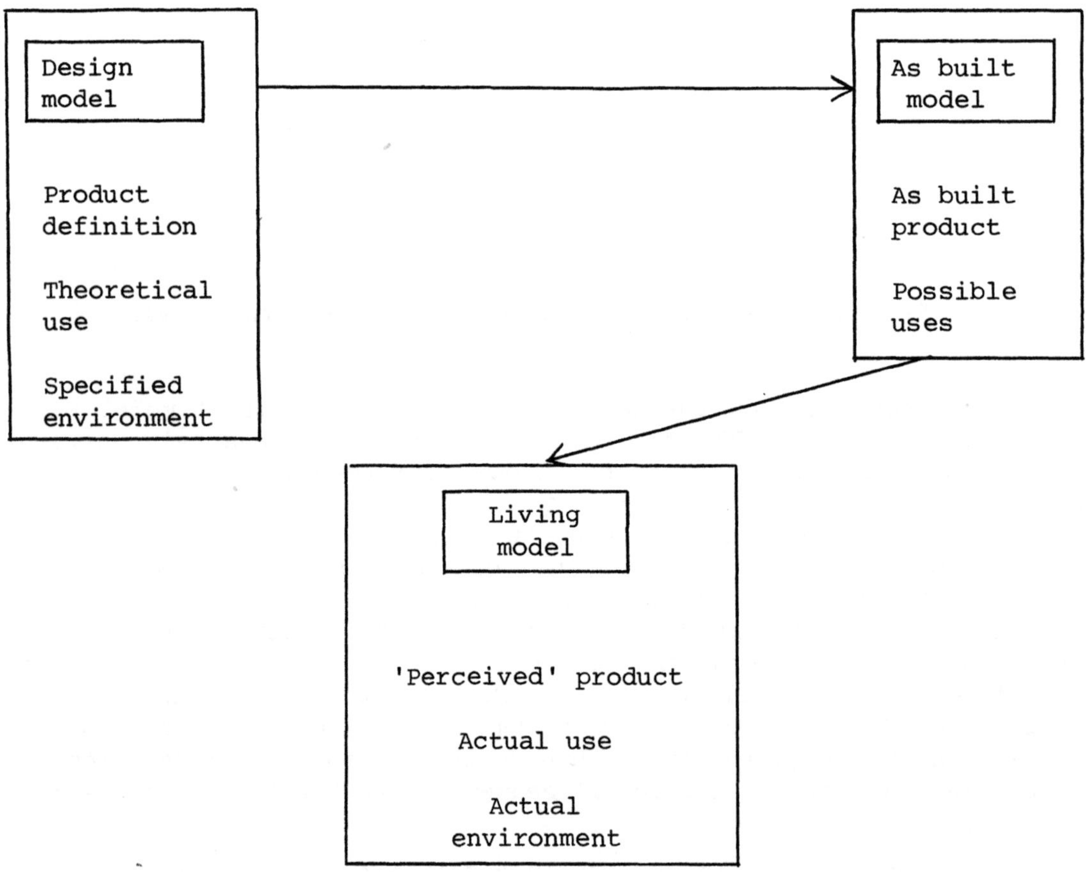

Fig. 5. The conflicting possibilities with respect to reliability

From this we can establish the following:

(1) A potential accident involves:
- the as built product
- its actual use
- its actual environment.

(2) The manufacturer can exercise influence on:
- the definition of the product
- the as built product
- its theoretical use
- its possible uses
- the specified environment.

(3) The user client can exercise influence on:
- the perceived product
- its actual use
- its actual environment.

In addition to these boundaries of influence, there are also boundaries of responsibility. The user may thus be held responsible for any accident originating in a modification made by him on the product supplied by the contractor. Conversely, the supplier may be held responsible for accidents caused by dangerous modes of use which are not prohibited by his user documentation.

We shall thus examine the position of the manufacturer and the user in succession, to identify the role which each can play in obtaining an acceptable safety level.

The Position of the Manufacturer

The manufacturer designs, develops and produces a technological product. The starting point for his work is a collection of requirement specifications either established by him following a market survey or imposed upon him by a call for bids. This is the document from which he defines his product. In general, user instructions are only prepared after the product has been defined. In addition, the requirement specifications are generally imprecise as to the list of environmental conditions under which the product is expected to function; it is generally limited to a specification of vague heat and humidity conditions.

The manufacturer is also aware of the possible differences between the designed product and the as built product. Companies with quality departments responsible for producing and applying inspection procedures on products under construction are becoming increasingly widespread. This function is, however, generally limited to a dual role:

(1) ensuring that the expected performance requirements are maintained throughout the production series;
(2) limiting products returned to the manufacturer by virtue of his legal or contractual guarantees to an acceptable rate.

This is the customary situation. From the scheme proposed earlier, we are able to present a typical structure of a reliability standard which any manufacturer should observe simply by virtue of his liability with respect to accidents potentially occurring on use of his marketed product. He is required, in this connection, to fulfil four principal tasks.

(1) He must be aware of the limits for the safe use of his product, and must inform his clients accordingly.

(2) He must define the limiting environmental conditions under which the product can be safely used, and must inform his clients accordingly.
(3) He must be in control of any differences between the designed product and the as built product which are liable to give rise to additional risks.
(4) He must research all possible uses of the as built product and take any technical measures required to prevent any uses which would be dangerous (e.g. simultaneous control of two contradictory functions).

Performance of these four tasks requires the production of a Safety Assurance Plan, comprising various activities performed throughout the development of the product, as shown by Fig.6.

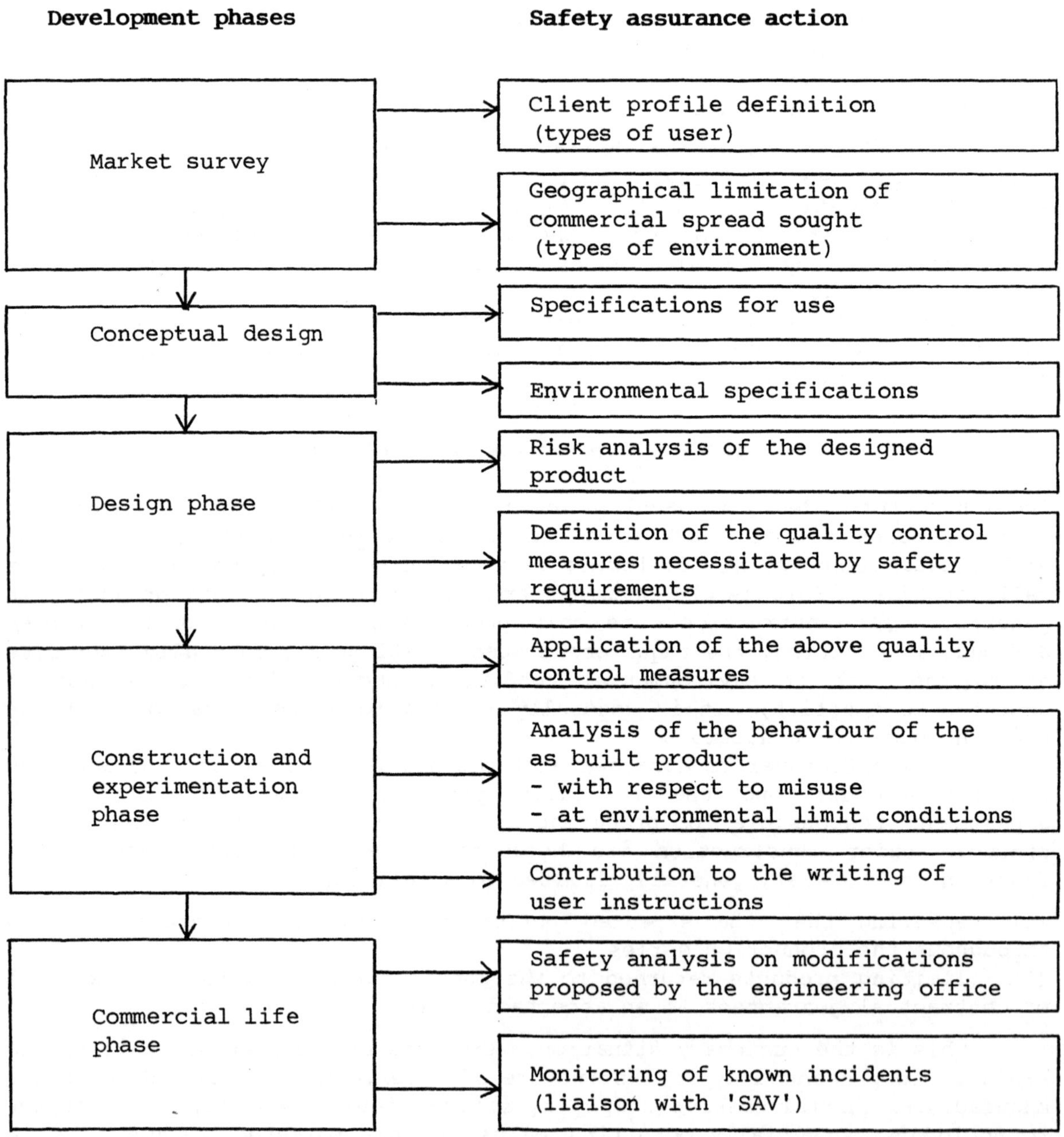

Fig. 6. Logic diagram used in the production of a Safety Assurance Plan

The Position of the User

The client is regarded here as the body responsible for a commercial or state organisation, purchasing an industrial product (e.g. a production plant). The risks which he runs as a result of making such purchase can be of various types:

- injury to user personnel;
- financial loss as a result of unavailability of production plant;
- damage to the organisation's property as a result, for example, of fire or explosion caused by the product purchased.

Some of these risks may be reduced (with regard, at least, to their financial consequences) by insurance or contractual protection vis-à-vis clients or suppliers. In other cases, this type of protection would be too costly (insurance), legally inconceivable (contractual protection) or humanly inconceivable (physical effects on personnel).

The client's essential task therefore lies in making a comparative analysis of the validity of the various conceivable means of protection in the face of each of the potential risks.

To sidetrack for a moment, if the law, through the notion of implicit legal guarantee, protects the client against hidden defects in a product, the client is not totally released from all liability with respect to the safety of the product which he uses or causes to be used. In 1982, a draft bill provided for possible stoppage, on union intervention, of a machine considered to be dangerous. In a case such as this, is the management of the company concerned able to fully insure for this type of risk? Can he transfer this liability to the supplier manufacturer, arguing on the basis of his good faith or ignorance when selecting the machine? It is probable that the client's liability with respect to safety will increasingly tend to begin at the moment of selecting the product. In addition, this selection itself is tending to become increasingly enlightened as a result of the control on advertising standards and the growing influence of user and consumer associations.

So what is the client's safety problem? In the most complex case, it may be broken down into six component parts:

(1) the choice of the safest product, taking into account the client's own use and environmental data;
(2) users to guarantee not to modify the product supplied by the manufacturer (or to carry out a precise safety analysis on any such modifications);
(3) users to be instructed on the product supplied, to improve their knowledge of the product itself, and the procedures for its use;
(4) the risks created by installing the product in its particular location to be analysed;
(5) choice of insurance policy;
(6) monitoring the product once installed, with respect to performance, but also to user's behaviour and possible developments in the natural and technological environment.

These various forms of action may be covered by a risk control strategy, drawn up immediately the decision to invest is taken. This strategy takes the form of a logic diagram, as shown in Fig.7 (and which, let us not forget, illustrates the most complex case).

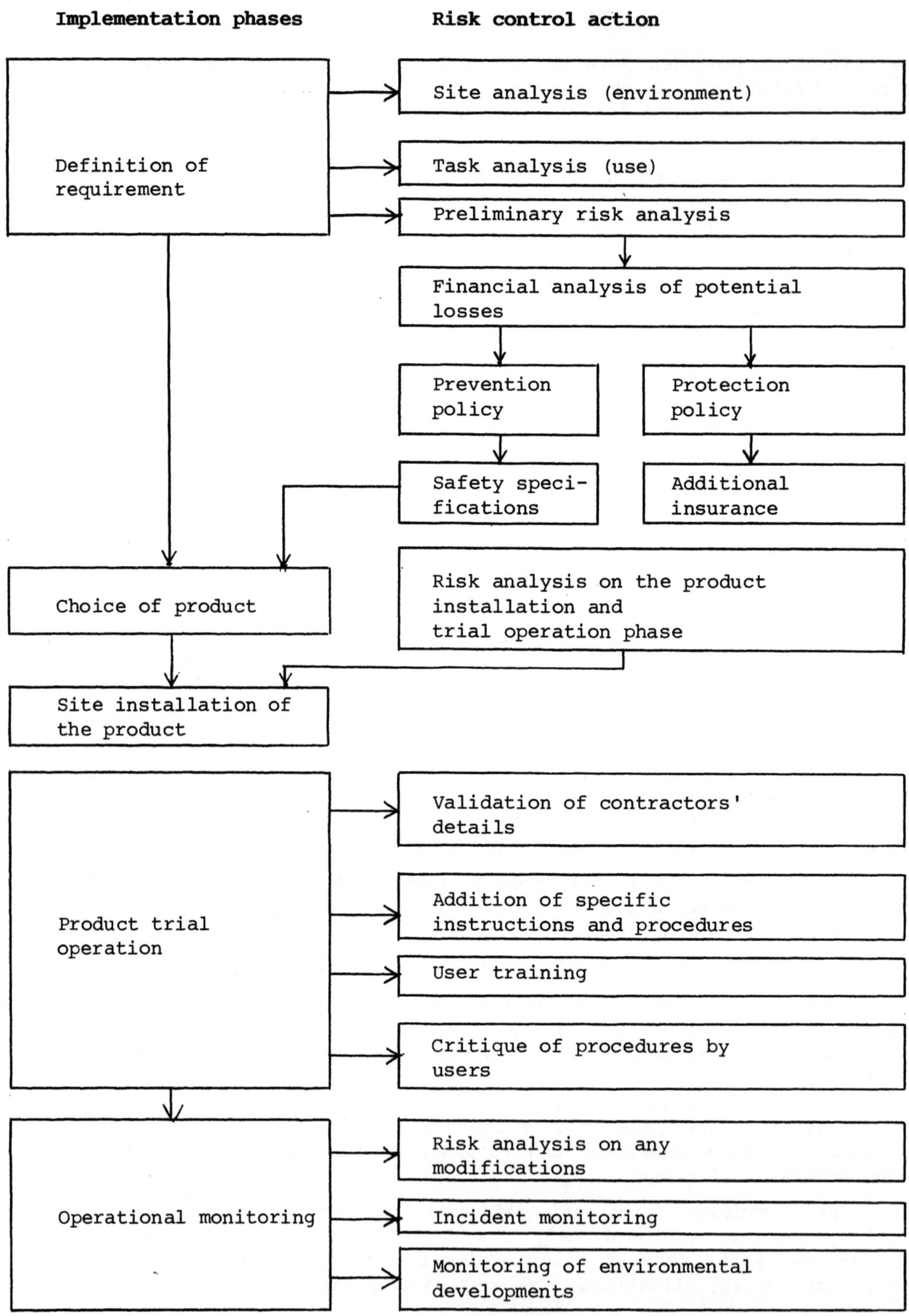

Fig. 7. Logic diagram for a risk control strategy

Dealing with Technological Risks

One conclusion is quite clear: the risks involved in the use of technological products, whether on the supplier's or on the client's side, cannot be dealt with either by the precepts of systems reliability or by risk management formulae alone. It would be unrealistic to seek to approach this question without reflecting on its legal implications and without taking into account the essential realities of environmental constraints and the multiple facets of human behaviour. Even on this latter level, new doors can be seen to open in relation to the safety ethic: does the safety of a social group inevitably have to represent a constraint on the individual, a restriction on his freedoms? Doesn't a lack of safety generate fear, hence alienation, etc.?

The questions are so numerous and so vast that to answer them is inevitably outside the scope and terms of reference of any one manager. But the questions are so numerous, too, that he is bound to perceive certain aspects in the daily pursuit of his activities.

We trust that we shall be excused for this philosophical digression at the end of these technical reflections, but it seems to us essential to the formulation of our conclusion: there is no magic formula by which a commercial manager can divest himself of his safety problem. There are, however (and we hope that we have taken a few steps in this direction) certain methodological guides for dealing with risks technically and financially, by assessing the relative 'weights' of safety parameters enabling decisions to be taken on priority choices. It is nonetheless true to say that if safety is a science, it is a human science, and that any decision taken on safety is such as to require psychological, even moral considerations to be taken into account.

Chapter 3

State-of-the-art and Procedures in the Nuclear, Aeronautical and Aerospace Fields - Classification of Systems Transferable *A priori* to Offshore Petroleum Operations

STATE-OF-THE-ART

The traditional approach to safety consists of applying rules prepared progressively after incidents or accidents.

If we wish:

(1) to take account of the safety of installations comprising new equipment items, where we have no significant historically based statistical data and/or where there is a large number of components (the systems notion) with interactions that are difficult to appreciate,
(2) to quantify the safety level (probabilistic notions) of 'conventional' installations,

we have to use new methods of analysis, i.e. risk analysis methods.

A comparison between the conventional methods of safety and those of risk analysis is provided by Table 2.

Table 2 Comparison of conventional methods of safety and those of risk analysis

Conventional methods	**Risk analysis**
Deterministic	Probabilities
'Pseudo zero risk' concept	Acceptable risk (non-zero) criterion
Oriented towards equipment having individual safety margins	Oriented towards systems
Based in part on previous experience (accidents)	Based on past accidents and on experience at the component level

Definitions

Risks. Magnitude in two dimensions associated with a precise phase in the life of the system and characterising an undesired event by:

- its probability of occurrence;
- the sum of the consequences resulting from the unexpected occurrence of the event.

Risk analysis. An analysis of risk can be defined by these objectives:

- identification of risks liable to occur in the operation of the system;
- study of the causes of the risks considered and, possibly, their quantification;
- calculation of the consequences of the occurrence of the risk in terms of the effects on human life, effects on the environment, economic losses, etc;
- proposing improvements to be made to critical events or weaknesses revealed, or to the system itself.

The Practice of Risk Analysis in the Nuclear Field: Risk Analysis Methodology at the CEA

Definition of the System Analysed and its Functions

Once a risk analysis study is initiated, generally in a somewhat vague manner, a proper description of the system is required. Before attempting to ascertain how it might fail, it is essential to understand how it works.

This requires a large amount of information to be collected: drawings, plans, expert explanations, procedures, etc. At this stage, flow diagrams are extremely useful for the purposes of handling and presenting the information to be taken into account.

Risk Identification

The next stage consists of identifying the potential risks exhibited by the system, particularly in the case of a new problem. There are a number of methods here, the most widely known being 'hazard analysis' (implemented within the framework of the Hazard and Operability Study method or HAZOP and 'preliminary (or elementary) risk analysis'.

The selected risk is governed by the stage of progress reached on the project, existing regulations, and the extent of any loss caused by occurrence of the risk.

System Models

When the system is too complex to be covered by a single block, it has to be broken down into a number of parts. This needs to be done carefully with a view to subsequent stages of the study. It is extremely useful to produce a model covering the overall behaviour of the system. Once again, a number of methods may be used: 'events trees' and 'cause and effect diagrams'. These result in a description taking into account both normal operation and failures in the system vis-à-vis the potential dangers identified in the previous stage. These methods naturally involve cutting down the overall system, identifying the various sub-studies required.

In addition, these global models can be re-used at the end of the study to combine the various results obtained in the part studies and evaluate the overall risk exhibited by the system.

Quantitative and Qualitative Analysis

Each part study also involves a number of stages. The first, of course, consists of a detailed description of the part in question. This is followed by an analysis of component failures: whilst this analysis is always performed, it may be explicit (on paper) or implicit (in the mind of the person producing the study). Analysis of failure modes and their effects of the system (FMEA - failure mode and effect analysis) is widely used for this purpose, and is sufficient for systems with a relatively low redundancy.

Even for the more complex systems, this method is extremely useful as a means of understanding how the system behaves when components are defective. Where FMEA will suffice, it is frequently supplemented by a 'criticality' analysis (becoming FMECA), or a quantification. Where it will not suffice, more powerful tools are used, leading to the production of a failure model for the system in question: reliability diagrams, fault trees, Markov diagrams, etc. From these models, it is possible to produce both qualitative and quantitative analyses. For systems with in-built redundancy, a study of common cause failures has to be carried out at this level.

All of the above methods are designed to reveal the weaknesses (critical paths) in the system in question; combined with a calculation of the consequences, they therefore lead to an evaluation of the potential risk inherent in the system. If a partial risk criterion is defined, this may be compared with the calculated risk, and if necessary, improvements may be made by working on the weaknesses.

When all part studies have been completed, their results must be combined with the aid of the model developed for the global system. Quantitative, qualitative and common cause failure analysis must be carried out here to evaluate the overall risk and to highlight the weaknesses (critical paths) in the overall system. If the overall criterion is not met, improvements must be made by working on the highlighted weaknesses.

It is important to note that the above description relates to risk analysis on a complex system for which a risk criterion has been defined and for which all necessary data (probabilities and consequences) are available or susceptible to evaluation. This, of course, is the ideal. In fact, more frequently, there is a lack of data at all levels: a lack of information, a lack of probabilistic data, no model from which to calculate the consequences, etc. As a result, the above ideal will not apply in full in all cases, and certain stages will have to be abandoned or reduced. The study may thus be purely qualitative, semi-quantitative or fully quantitative, and its object may be a precise evaluation of risk, or simply a relative evaluation, to enable a choice to be made between various designs, etc.

Computational Codes

The analysis and processing of large fault trees (or of any other method of assessing systems reliability/safety) requires the use of computational codes.

Codes have been developed for processing:

- fault trees - the CEA's LOCK and PATRICK, EDF's DEFAIL and FIAB C;
- marker graphs - the CEA's MARK-GE, and EDF's FIABI;
- also, the CEA has developed MONARC and BIAF codes to carry out Monte Carlo simulations.

Reliability Data

Since all risk analysis is carried out with a view to quantification, it has been necessary to develop data banks to provide representative data back-up for these calculations.

The EDF, for example, has set up a reliability data bank (SRDF - Service de Recueil de Données de Fiabilité), covering the six PWR sections of Fassenheim and Bugey. The SRDF monitors 800 items of equipment (both electrical and mechanical) by section.

FEATURES SPECIFIC TO THE PETROLEUM INDUSTRY

The features specific to the petroleum industry as compared with the nuclear industry are:

- Time and sequence interdependence. The systems studied comprise a series of operations, with the assembly of a certain number of equipment items. Thus, for offshore operations, an analysis of the level of meteorological risk has to be taken into account at each stage, this risk being directly dependent on the time required to perform the operation, and the date on which it is carried out.

- An absence of reliability data. This problem was also encountered when producing the first risk analyses in the nuclear field. Data banks or data bases have been produced as and when the need has arisen.

- A predominant level of manual work, whether in drilling (on or offshore) or in driving operations.

- The need to contain all equipment required for the due operation of an offshore unit within a restricted space.

TRANSFERABLE KNOWLEDGE

Methods

All of the specified methods and methodology can be used by the petroleum industry. With other methods currently being developed (such as Petri networks) it should prove possible to cope with all problems involved in safety and reliability assessment of petroleum systems.

It is important to note, however, that the major problem lies in the way these methods are used and, in particular, in the way in which they are combined to study a particular problem. This problem can only be resolved by producing general risk analyses.

Computational Codes

Existing computational codes can be used directly by petroleum operators. They simply require synthesising to enable a rapid choice to be made of the appropriate computational code for the problem in question.

Part II

The Application of Risk Analysis Methods to New Offshore Oil Structures and Operations

Chapter 4

The Possibilities of Applying Safety-Reliability Methods to Offshore Oil Operations as Provided by Community Regulations (Regulation 3056)

INTRODUCTION

As we have seen, safety-reliability procedures using fault tree methods of risk analysis are used for mechanical or electronic systems, nuclear installations, chemical complexes and aerospace operations. They are not, as yet, widely used in the field of hydrocarbon exploration and production, whether onshore or offshore. There have been instances of particular applications, e.g., for deep-sea exploration risers. These safety-reliability techniques have been used in the past, and continue to be used, for entire fields (e.g. in particular the economics of a marginal field, SNEA (P)) and for the optimum selection of component parts of a deck (modules) of a production platform (CFP).

Their more widespread use would at first appear likely to relate to entire fields, from the dual aspect of capital expenditure and safety, but, in fact, their development is concentrated on structures, where analysis is tending to be based more and more on a probabilistic approach, as opposed to the deterministic philosophy of the design codes. This application to structures would enable the risk of structural ruin to be explicitly analysed, and would constitute a logical and practical extension to the advanced procedures for structural design. This development is not only desirable but foreseeable, and its pursuance is highly recommended in order not to lose out to competition.

TOWARDS A NEW STRUCTURAL DESIGN

Blow-out Control and Prevention

Any blow-out control and prevention plan must in part pass through risk analysis. The object of this analysis will be to highlight the prevailing accidental sequences leading to a blow-out, and hence to bring the blow-out under control once it has occurred. However, these various techniques will be to no avail if operators are not regularly trained on blow-out simulators.

Special Equipment

The special equipment such as the dynamic positioning system for drill ships or the control-command system of an RCP could be easily subjected to an exhaustive reliability study. Work has already been carried out in this connection, particularly in Great Britain and Norway.

Production

Installation Availability

Representative reliability data (including mean times to repair) for offshore conditions are beginning to appear and it is therefore conceivable for reliability and safety studies to be carried out on an offshore production unit. These studies would be designed to:

- ensure that the unit in question was at least as reliable as units already in existence;
- determine the mean availability of the unit, allowing production scheduling for future years (economic consequences of the technical risk);
- optimise stocks of spares.

New Concepts

The level of risk for new production plant concepts may be calculated with respect to both safety and economy, at least in its broad outlines.

Particular reference may be made here of the SKULD experimental testing station, its engineering being based largely on the probabilistic concepts of reliability. From the very beginning of the exercise, the emphasis was laid on the safety and reliability of the station, enabling choices or compromises to be made on the basis of stringent technical criteria.

This type of analysis, even if only qualitative, could be usefully applied to other areas. But the principal danger lies in the temptation at all costs to set probabilities for aspects which are simply not susceptible to evaluation at the conceptual design stage.

Safety Systems

To avoid freezing the design of safety systems at a given point in the history of the relevant technology, it would be a good idea for comparative reliability studies to be carried out on 'conventional' system concepts as compared with the 'innovative' system. These comparative studies could:

- partially eliminate the uncertainty or poor quality of data;
- do away with the need to define the acceptable risk level (i.e. the minimum reliability of the system).

It would similarly be useful to define the frequency of testing of safety systems on the basis of reliability studies. Indeed by definition these systems are on stand-by, and faults can occur without such an occurrence being visible. This would eliminate the dilemma of:

- frequent testing, where the state of the components is known, but equipment suffers from the tests;
- infrequent tests, where the state of components is therefore unknown (concealed unavailability).

Figure 8 illustrates the problem: two testing frequencies T_1 and T_2 relate to the same unavailability I_o (i.e. the same safety level): T_2 is selected as the more economical.

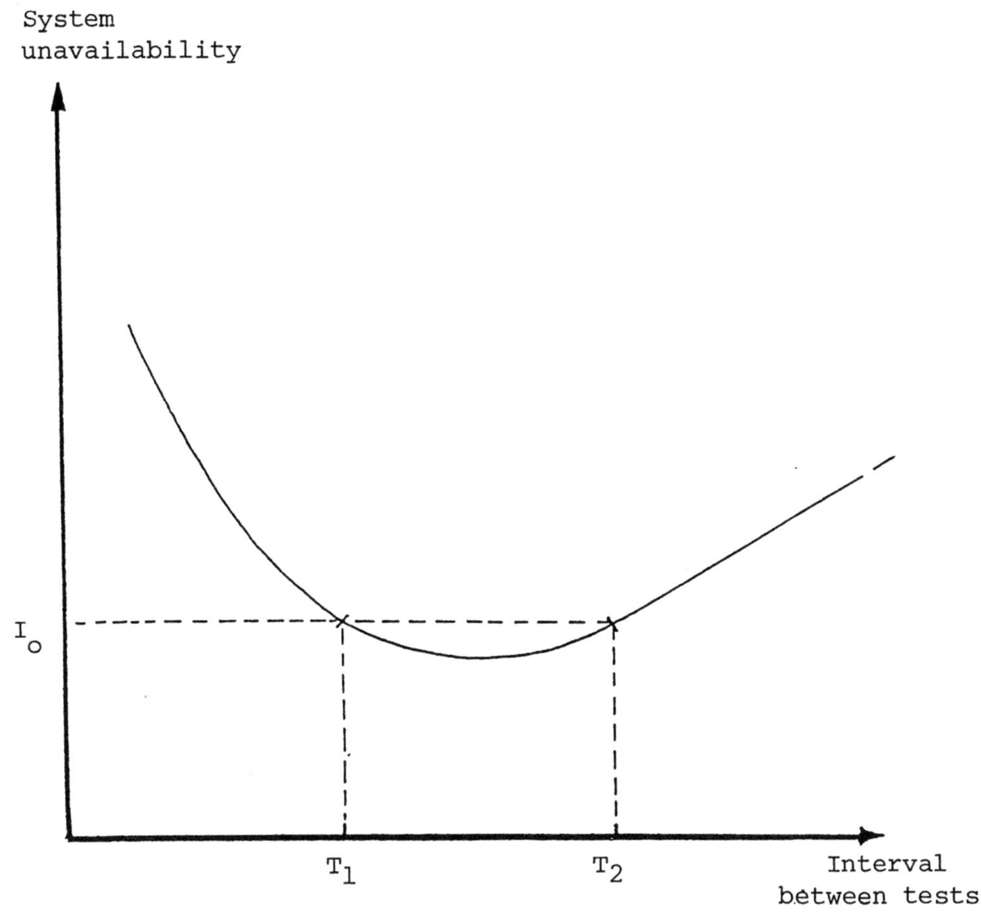

Fig. 8. Testing frequencies

Control Room Design

In the same way as the premises of a drilling unit, the behaviour of control rooms in the event of explosion could be calculated, the object (as in the chemical industry) being to define a control station capable of withstanding a catastrophe and which could still be controlled and possibly made safe.

However, the most important point to be studied is that of the actual design of the control-command unit, and not its structure. This involves the path recently followed by EDF in permanently supplying the operator with aids to decision-making, by means of elementary data processing. The Three Mile Island incident clearly revealed this need for a more in-depth analysis of the system, not simply supplying the operator with a pre-processed set of information.

Product Despatch

Loading Stations

Here, the development of new systems is based on the use of new components (rotating joints, cryogenic hoses). The only possible way of assessing the quality of these components is by (1) studying their reliability (mechanical reliability); and (2) above all by carrying out reliability tests (which may or may not be accelerated). This involves testing a number of items to destruction, if possible. However, this is extremely costly and can only be carried out on a limited number of items.

Pipelines

Inspection. Availability studies of the type already produced can enable an inspection and maintenance policy to be clearly defined for pipelines in service.

Repairs. Reliability studies have already been successfully carried out within the framework of the project under the direction of Total-CFP and SNEA(P), in association with ACB and Comex. These were designed to prove the reliability of an automatic repair system for deep-sea pipelines. Qualitative risk analysis was carried out on the failure of offshore testing of the project (FMECA and fault trees) and was felt by all parties involved to have been highly satisfactory. It enabled:

- a number of unnecessary redundancies to be eliminated;
- the quality of the system to be improved by eliminating weaknesses;
- implementation procedures to be clarified, including those of the surface operators.

CONCLUSIONS

There are two ways of assessing project safety and reliability levels, depending on whether or not an overall risk criterion has been defined.

The Use of Probabilistic Methods Without Global Risk Criteria

Analysis is carried out in two stages:

(1) A full risk evaluation related to a reference unit: this evaluation may be derived from an exhaustive risk analysis, or from statistics.
(2) Relative evaluations in relation to this reference, to eliminate (a) criteria definition and (b) some of the uncertainties.

It would then be possible for existing units:

- to assess the significance of any modification;
- to establish procedures in the event of partial unavailability (particularly of certain items of safety equipment);

and for future units to provide a design aid by allocating risk between the various sub-systems.

The major disadvantage of this approach lies in the diverse nature of oil installations.

Definition of an Acceptable Risk Criterion

Criteria are currently being established by the 'Nuclear Regulatory Commission' in the United States (see Appendix 2). There is, however, a probabilistic criterion already in use in the offshore oil industry. This relates to 'Guidelines for safety evaluation of platform conceptual design' by Norwegian Petroleum Design. The content of these guidelines is summarised below.

Stage 1. Definition of the accident considered (e.g., surge pressure in a pipeline).

Stage 2. Calculation of the annual probability of occurrence of such an accident.

Stage 3. Analysis of the consequences of the accident (e.g., calculation of any surge pressures (and associated probabilities) caused by the occurrence of such an incident).

Stage 4. The assumed consequence (e.g., the surge pressure wave value) is determined if the cumulative total of probabilities of occurrence of the consequences, based on the highest value (i.e. surge pressure waves of 2-1.8 bars) reaches a value of 10^{-4} per year (e.g., the assumed consequence is a surge pressure of 0.2 bar).

Stage 5. Calculation of the damage caused by this consequence.

Stage 6. Check that the damage still allows:

- at least one escape path to remain intact for at least one hour;
- refuge zones to remain intact until evacuation can be undertaken;
- the supporting structure to continue to fulfil its function.

Chapter 5

Drilling Activities

DRILLING FACILITIES: FIXED AND MOBILE PLATFORMS, DRILLSHIPS

For drilling facilities, probabilistic calculations are carried out on ultimate stress and fatigue strength to determine the sample checks required (section and modulus of inertia of structural members) to ensure the capacity of the material used does not exceed at any time the anticipated life of the structure. External influences, particularly wave action, are regarded as random events. Development is more advanced for these influences (anticipated use), although considerable additional data and physical/ mathematical interpretations are still required, than for the complementary area of strength (capacity), where practical values (the variability of material characteristics and manufacturing defects) are still frequently missing.

Fatigue, one of the engineer's main concerns for many years, raises practical problems. It has however been the subject of a vast European programme started six years ago. The use of the results of this programme follows the straight line of the probabilistic methods.

One particularly important aspect in structural safety is that of structural 'redundancy'. Redundancy is the property whereby a number of structural members have to be destroyed before the entire structure fails. Unlike mechanical parts, all structures designed by the engineer have in-built redundancy, often as a result not of analysis but of common sense.

Redundancy, explicitly studied and introduced into reliability and safety studies in the aeronautical and nuclear industries on electronic and mechanical systems, has not been studied or introduced in marine structures, though the need has often been stressed*.

Today, modern procedures are being applied more to platforms than to ships. This is explained by the difference in architecture (the former comprising a skeleton framework and the latter a body and hull, distributing the external stresses) and by the considerable difference in experience in the use of these two categories of marine facility. Modern procedures are, however, being used on unconventional vessels such as methane tankers and container ships. They are scheduled to be used for the projected ice-breaker methane tankers for the development of resources in the Canadian Arctic. On

---

* Cf., in particular P.W. Marshall, 'Les compromis entre risque et coût dans le projet et le contrôle de fracture des plateformes offshore' (Compromise between risk and cost in the design and control of fracture in offshore platforms), Technical Bulletin issued by Bureau Veritas, June/July 1980, and J.P. Planeix and O. Ostergaard - 'Procédures et stratégies modernes pour le projet des structures marines' (Modern procedures and strategies for the design of marine structures), Technical Bulletin issued by Bureau Veritas, September 1982.

the other hand, risk studies which could serve as a model for other applications, have been carried out in the case of appreciable risk shipping operations[+].

DRILLING TECHNIQUES

Automation of Operations Performed on the Drilling Floor[#]

Work on the drilling floor has traditionally been tough. In the open air with a minimum of protection against the weather and using crude equipment, for many years it has been a job for strong arms and rough hands, particularly with onshore rigs. It is only recently, and essentially on offshore rigs, that a gradual evolution has begun to take place in the mechanisation of handling activities (including rig operation), which represent a major proportion of the work (approximately 30% of working time). Rig operating devices have begun to appear, particularly on floating facilities - power tongs and various mechanical handling aids.

On dynamically positioned ships, given the various responsibilities of the drilling foreman and the fatigue inherent in 12 hours of continuous shift-working, an attempt has been made to provide the foreman with a comfortable environment, placing all the controls inside an ergonomically designed cabin. This concept tends towards a more generalised use of remote controlled devices.

At the same time, the trend towards mechanisation has seen the introduction of equipment for mechanised rig operation (the 'iron roughneck'). More recently, a number of rigs have been fitted with high power motorised swivels (800-1000 CV) replacing the traditional rotary table in the floor. This innovation provides an improvement in safety, enabling mud to be circulated and drill pipes to continue rotating during manoeuvring. When combined with other existing mechanical devices, it should reduce the arduous nature and risks of working on the drilling floor.

Whilst drilling is unlikely, in the immediate future, to acquire the sophisticated, even intellectual character of geophysics, it is making increased use of the resources offered by the new technologies: data processing and robotics. The former, by processing an increased number of qualified measurements, can increase the content and information yield of drilled wells. The latter should transform the working conditions of drilling personnel. The combination of the two will undoubtedly bring a qualitative revolution in the relationship between drillers and their work. This should attract individuals with a higher level of education than in the past, and should accelerate a trend which has already begun to manifest itself.

The Introduction of Robotics on the Drilling Site (see Appendix 5)

The essential aims are:

- to improve manpower safety and, as far as possible, eliminate the most dangerous jobs;
- to reduce the physical fatigue of personnel;

[+] Cf., in particular J.P. Jaunet and Y. Le Gal, 'Trafic maritime et derives accidentelles en Manche' (Maritime traffic and accidental drifting in the Channel), Technical Bulletin issued by Bureau Veritas, September 1982.

[#] For details on this subject see Appendixes 5-15.

- to reduce unproductive time;
- to ensure regularity of handling operations.

New equipment items and appropriate methods of use need to be investigated with a view to progressive automation of drilling rigs. The starting point may be new items such as automatic slips positioners, drill pipe and collar spinners, rotary drive swivel heads and casing handling systems. It is essential for new equipment items to be designed, if a presence on the international drilling rig market is to be established.

Proposals

The plan of action could involve two phases.

Phase 1 - preliminary phase

- Study of current procedures: breakdown of operations performed on the drilling floor into elementary operations, analysed from the viewpoints of manpower, equipment, duration, safety, frequency, etc.
- Preparation of an inventory of existing equipment, visiting drilling rigs and manufacturers.

Points requiring particular study are as follows:

- fitting and removing drill collars;
- fitting and removing stabilisers and drilling turbines;
- pipe handling in round trips and drilling;
- handling of casings;
- handling of heavy equipment items (e.g., injection spools, cementing heads, wellheads);
- making-up, breaking-down, tightening and breaking-out drill pipes, drill collars, casings, with a particular study on the power swivel.

Phase 2 - development of new procedures and specifications for new equipment

From the results of Phase 1:

- identification of procedures requiring improvement either by amendment or rewrite;
- writing specifications for new or modified equipment;
- also the determination of procedures for the use of equipment under development (making-up/breaking-down, tonging, power swivel, casing handling equipment, etc.) in conjunction with a derrick or conventional mast, to achieve the maximum efficiency. This may also result in specifications for new or modified equipment.

Risks in Offshore Drilling Activities and Control of Operations: Safety Codes and Procedures

The number one risk in drilling (disregarding the risk associated with the naval drilling facility) is blow-out. Since the drilling industry was first organised, the notions of risk analysis have been applied to drilling operations and systems, whether on a rudimentary or more advanced scale.

Attempts have been made to anticipate the consequences of any event in a given operation, whether routine or otherwise, with a view to issuing advance instructions as to any action to be taken in the face of such an event, to avoid any harmful consequence. This type of approach leads to the establishment of operational procedures. In the case of systems (operational combinations), it leads to a safer design of equipment, sometimes incorporating duplicated systems (multiple control points, dual circuits, etc.). In all cases the approach can be assisted by risk calculation, and this is becoming increasingly common.

Following a brief general discussion of the subject, we shall provide an illustration, (without calculations) of a study, on the one hand, of the risk of blow-out during the course of drilling operations, and, on the other, on the design of wellhead equipment, with their resultant safety codes. Further details are given in Appendix 16.

General Approach

Preventive phase. There are two aspects to the preventive phase: risk assessment and the establishment of preventive procedures.

Risk assessment involves:

- definition of the undesirable event(s);
- identification and analysis of its (their) cause(s);
- identification of the immediate consequences or easily detectable indicators preceding the undesirable event.

Preventive procedures should be implemented when any of these 'indicators' is observed. These procedures are to be based on:

- the inventory of systems installed, and their limitations;
- definition of the responsibilities of each person involved.

Corrective phase. The undesirable event has occurred. The corrective phase has two levels: (1) action using available resources, and return to the normal situation as applying prior to the undesirable event; and (2) an escalation in the seriousness of events, following the failure of corrective action. It is impossible to return to a normal situation without recourse to external resources.

In both cases, but more particularly in the second, the corrective or control operations will require:

- a pre-existing emergency organisation, known to those involved and in charge;
- a list of potentially useful and readily available resources;
- selection of a control method based on experience in past events;
- the organisation of control operations based on fault trees broken down into individual operations and translated into operational procedures at site level.

Drilling Equipment Reliability

Design phase. At the equipment design stage, risk analysis is used to highlight weaknesses in the system in the light of its conceptual design (which can be modified) and its operating conditions (which cannot be changed).

If the risk of failure of equipment under normal operating conditions is high, its conceptual design should be revised to reduce this risk to an acceptable level.

If the risk of failure is high only under extreme operating conditions, it can sometimes be reduced by duplicating weak systems or setting-up an operating procedure which avoids exposing the equipment to extreme conditions.

Risk evaluation at the design phase is to include the risk of failure as a result of the long-term use of the equipment, which then becomes subject to fatigue. Fatigue and operating conditions may be combined, the operating conditions of a fatigued item in fact being liable to constitute extreme conditions.

Testing phase. Once the equipment has been built, it has to be tested under normal and extreme operating conditions to assign design strengths and locate its potential range of use. Depending on the gravity of the function to be performed by the equipment, these tests may be carried out on a sample of systems manufactured, or on each individual system.

Incidence of the integration of a given item of equipment in a system on system failure risks. In general, this involves applying risk analysis to a system comprising a number of equipment items, in order to ascertain the usefulness of attaching an item of equipment to this system intended to make it at least applicable, if not more reliable, under extreme conditions, in comparison with those for which it was designed. The best example is the risk study carried out at CFP on the BOP and injection spool wellhead.

Safety Codes - Practical Exercises and Tests

Safety codes. All safety codes derive from the concern to limit risks. They are the product of advanced or rudimentary analysis of the risks involved in a given operation and in the systems used. The most difficult aspect to take into account is the human factor, but safety codes affect the men who will be applying them, and they must, therefore, be sufficiently restrictive at least to limit this factor in operation. The examples illustrate the progression from analysis through to the code (see Appendixes 5-15).

Practical exercises. Practical exercises are the best means of testing the reaction of emergency teams to undesirable events. Exercises necessarily must follow on from training on the risk in question, teaching operatives the reactions required in the preventive and corrective phases, and also indicating the causes of the undesirable event, and how to avoid them. Practical exercises must be carried out at regular intervals, and should be followed by a critique. Simulators of the main risk operations in drilling are now available: these should be of assistance in manpower training and qualification phases. Simulators cannot, however, totally replace practical exercises in the field.

Testing. This relates to periodic testing of the various critical equipment items. This testing must provide a check on the ability to function correctly when required. In certain cases, this testing can introduce a certain degree of fatigue to the equipment, and this must be taken into account. This procedure may form the subject of risk analysis.

For further details on blow-out control and prevention see Appendixes 15 and 16, which describe the techniques currently envisaged to prevent blow-out and the procedures to be followed in the event of blow-out.

Chapter 6

Production and Processing Operations

NEW PRODUCTION RIGS

As far as analysis of the method of calculating external stresses is concerned, there is no difference between the conventional production rigs (whether, as in most cases, fixed (jacket type, gravity platforms) or, more rarely, mobile but held in position for long periods of time for production purposes (jack-ups and semi-submersibles)) and the corresponding drilling rigs: these units are subject to considerably greater stresses in given wave directions than are mobile platforms operating successively in different locations. Mobile platforms used as 'floatels' may be considered here under the same category as mobile production platforms. Only advanced probabilistic procedures can take into account the difference in effects between these two types of use.

For newer fixed types of production rig (tension leg platforms, guyed towers, articulated gravity platforms and multiple articulation platforms) modern procedures are particularly recommended, enabling an evaluation of the influence of particular elements in the structure (e.g., the anchor legs of tension leg platforms) on the safety of the structure.

PRODUCTION DRILLING

The question of safety in production drilling is dealt with in conjunction with safety in exploratory drilling (see 'Drilling techniques' above), since in this respect there is no difference in nature between these two activities. Indeed, in both cases, the essential risk is that of blow-out.

LIVING QUARTERS

Work on living quarters has concentrated on an essential safety aspect, namely platform evacuation, where the development of new systems is urgently required. Experience has shown that the current systems of offshore evacuation do not properly fulfil their objectives. Evacuation capsules are only reliable in seas of a maximum force of 6-7, while platforms are designed for conditions which are frequently extreme (wave heights of 30-32 m, 100 knot winds).

Improvement in Safety on Platforms

Between the years 1955 and 1982, 40 drilling platforms have been lost and 32 severely damaged. Under extremely unfavourable weather conditions, the loss of stability due to problems of floating structures can involve the total loss of an offshore platform plus its crew. In such cases, rescue apparatus must exhibit an extremely high level of marine resistance. Unfortunately, current life systems are inadequate (Appendix 17). This is a well-known fact and even tends to aggravate problems, since platform managers will hesitate to use them and a manager will only give an order to evacuate if he is quite certain that his platform is about to be destroyed. The total evacuation time is thus very short: 5-10 minutes for jack-ups; and 10-15 minutes for semi-submersibles.

In addition, with oil development in sub-Arctic zones, survival conditions will worsen. The A. Kielland accident killed 123 people, and the Ocean Ranger accident 82. The use of semi-submersibles and jack-ups for production purposes will give rise to further problems.

A Technical/Economic Approach

Following figures such as these, it may seem cynical to refer to economy in safety systems. It is, however, a reality, and one which we have to take into account.

Current systems are not expensive in comparison with the cost of the platform for whose personnel they are intended to offer ultimate safety: approximately $300 000 for a 50-man capsule with a launching system, i.e. in total, less than 1% of the cost of a drilling platform, and much less for a production platform. This figure is very low when we consider the other safety costs (oversizing of structures, fire-fighting systems, etc.). Indeed, the essential part of expenditure on safety on an offshore platform is aimed at preventing the destruction and abandonment of a platform - a risk which will always exist.

Possibility of Introduction of a New Concept

Evacuation constraints vary according to the types of platform (freeboard, floating or otherwise, number of personnel, etc.). The introduction of a new evacuation concept onto an existing platform similarly gives rise to design and operational problems.

Appendix 17 provides a report on the various currently conceivable formulae for evacuation.

FLARES AND THE SAFETY OF PERSONNEL

The presence of flares, an essential item in the safety of production equipment, is a source of evident nuisance for personnel. This nuisance can usefully be analysed, therefore, with particular reference to radiation, to study the means of protection at various levels:

- studies on simple and low-cost flare supports providing the possibility of satellite location, away from production platforms;
- studies on protection systems for personnel: screens, special clothing, etc.

The Influence of Wind on Radiation

Theoretical studies are essential. These must include small-scale experimental simulation tests, to allow the incorporation of assumptions in the model. It would appear that:

- the development of flux radiation to the ground cannot be covered solely by aerodynamic parameters characteristic of the fuel and air flows;

- extrapolation of the radiated flux requires a local mathematical model which must be coherent and sufficiently complete to describe the interaction of flow, combustion and radiation in a three dimensional configuration;
- the relative increase in the flux radiated to the ground may be as high as 500% at laboratory level.

This results in a cumbersome model requiring considerable time to calculate, as the model is complete, with very few assumptions.

Performance of Field Tests

Tests must be carried out as a back-up to the model. These tests, therefore, have to be designed in such a way as to show and characterise all fluctuations of more than 1 sec. in length of time, and of potentially major importance for the human body.

The basic tests should consist of measuring the field of radiation (as a function of the distance from the flare) for 2 in., 6 in. and 10 in. tubular flares, with two wind velocities at flow rates of 100, 200, 400 and 800 m^3/day of cleaned gas. All physical quantities should be recorded simultaneously. Measurements should then be reconciled with the design results.

Other tests should also be carried out to test particular 'tips', quantifying their radiation in comparison with the model, together with noise measurements and measurements of radiation attenuation by screening.

Human Tolerance of Radiation

Investigations to test the validity of API standards for human tolerance of radiation have shown the following.

- The API values of 440 BTU/ft^2 x h^2 for continuous exposure and 1500-2000 BTU/ft^2 x h^2 for emergency flaring with an exposure time limited to a few seconds are valid (precision being difficult to assess, at around 20%) for an average individual whose skin is exposed (with particular reference to the top of the head).
- Tolerance varies considerably from one individual to another: e.g., from 20 sec. to 2 minutes, but with the possibility of establishing a mean.
- Clothing is to play an important part: a helmet enables the wearer to move away without risk or pain at a radiation of 2000 BTU/ft^2 x h^2, while bare-headed individuals are seen to flee in pain. A worksuit enables the wearer to withstand the maximum flux without discomfort, the body being protected. The role of ventilation seems to be negligible.

In view of the above, it would seem pointless to pursue a study of human behaviour with respect to radiation. The API figures may be adopted, but the importance of clothing (helmets, gloves and overalls) must be stressed.

THE SAFETY OF OFFSHORE TREATMENT PLATFORMS

The design of treatment platforms is based firstly on a number of safety considerations aimed at protecting human life, preventing pollution risks and safeguarding investments. For a number of years, the application of risk analysis to oil structures has also provided a more efficient means of dealing with the safety problem on offshore installations. The principal safety considerations and their implementation applying to the design of offshore production platforms are examined here.

The first concern relates to an efficient network of personnel evacuation routes from the various operating zones to the living quarters and to the lifeboats in cases of emergency. The locations of these items are decided by criteria of distribution among the various zones classified according to the probability of presence of flammable gases, i.e. the wellhead zone, the zone containing non-fired production equipment, the petroleum product storage zone, the fired equipment zone, the rotating machinery zone, and finally the living quarters.

The major concern in this zoning is to avoid flame sources in zones where gas is liable to be present. This classification has a direct effect on the design and implementation of ventilation systems, both natural and artificial, the pressurisation system, and also the type of electrical equipment used.

The platform is further equipped with an emergency shutdown system, causing either complete shutdown of the production facilities or shutdown of part of the facilities only, depending on the degree of seriousness. Flare systems, in particular, are designed and located as a function of the quantity of effluent to be discharged, the prevailing wind direction and the location of other equipment items.

The platform is also provided with a fire protection system comprising various devices for fire detection, isolation of flammable zones, air intake shutdown, and flame extinguishing. The principal structural members are generally protected so as to withstand direct exposure to flames for at least two hours without loss of mechanical strength.

As far as operating processes are concerned, operating sequences are established for all separation and treatment operations, and also for production plant repair and maintenance operations, with particular reference to wireline, workover and well cleaning operations.

Appendix 26 deals in greater detail with the subject of the safety problem in offshore treatment.

NEW CONCEPTS

Since the technique has to precede the need, the development and subsequent use of new concepts in increasingly sophisticated areas to satisfy operators is one of the constant concerns of the oil industry. In many cases, however, new concepts derive from previous projects and experience, and are less than revolutionary. This is the case here, but it is essential for totally original solutions to be found in the long-term future.

The Quest for New Concepts in the Development of Offshore Oilfields

The search for new hydrocarbon deposits offshore has led oil companies to require new techniques and equipment for exploratory drilling in increasingly deep water. Numerous floating facilities are now able to drill safely through 300 m of water. Some operate in sections of 1000 m, some even in 1800 m of water. On the other hand, the systems needed to bring in offshore deposits have only so far been applied to water depths of not more than 100-150 m, with a few exceptions of development for greater depths (e.g., the Cognac Tower). To work reserves in depths of over 150 m, new concepts are required, in three particular areas:

(1) support facilities (for production and assistance);
(2) subsea equipment (wellheads and subsea manifolds);
(3) emergency and maintenance systems.

The extrapolation of current techniques meets with technical and economic limitations, given the increased weight of structures and difficulties in inspection, maintenance and troubleshooting on pipelines and subsea equipment.

In developing these new concepts, the limitation of costs is bound to be a matter of major concern with respect both to the cost of the equipment and installation, and the level of maintenance costs. Costs must not, however, be limited to the detriment of production reliability or safety. These two aspects, reliability and safety, add a complexity to any development of a new concept.

In recent years, numerous experimental projects aimed at testing new concepts or equipment for deep-water duty have been carried out and installed in shallow water. For example:

- subsea equipment and maintenance/trouble-shooting systems: Exxon's SPS, Elf Aquitaine's Grondin station, Shell Esso's UMC;
- platforms: Exxon's guyed tower, Conoco's TLP.

The Choice of Platform for Deep-water Locations

Table 1 in Appendix 18 sets out the respective advantages and disadvantages of the main types of platform available to date for deep-water applications in the North Sea (300 m). These platforms may be classified according to type of platform and associated production type (see Table 3).

Table 3. Classification according to platform and production type

Type of production platform	**Rigid**	**Flexible**
Surface	Piled steel jacket Concrete gravity unit Steel gravity platform Steel tripod tower	TLP Guyed tower Articulated tower
Submersible		Catenary anchored floater

Depending on environmental conditions, the limiting depths for the use of rigid platforms are around 300-400 m. Beyond this depth, flexible-type platforms must be used.

For reasons of reliability and safety, platforms enabling production platforms to be raised to the surface are currently preferred, and will probably continue to be so. In effect, this limits the cost of maintenance and ensures improved reliability. A choice such as this does not, however, exclude the use of subsea production modules when the configuration of the reserve is such that all wells cannot be drilled from a single site.

Tension Leg Platforms (TLPs)

The various designs studied by the oil companies.

Amoco

Amoco appear to have been the initiators of the TLP concept. In this initial design, the anchor legs also serve as drilling risers: the wells are thus drilled through these legs. The production modules are located on the surface of each of the four platform risers (see Fig.A.18.7).

Gulf

In developing their version of the TLP, Gulf changed the philosophy. Their initial design used a six-leg hexagonal platform. Wells were drilled through the six legs on piled templates, which acted as drilling guides and an anchoring base. In a second attempt, this design was rejected due to the risk of impact between risers: it would appear that an eight-leg square symmetrical platform was adopted, the number of risers being minimised and production being centralised at a subsea manifold, at the platform centre vertical.

Chevron

Chevron's design is a symmetrical, square four-leg platform: the wells are drilled through two templates located across one of the neutral axes of the platform: the production modules are located at the surface.

BP

BP's TLP was a four-leg square platform on a single central template. Work on this type of rig, known as the TBP (tethered buoyant platform) was stopped on grounds of 'the absence of potential applications for this type of platform'.

Conoco

The TLP designed by Conoco and built on the experimental basis for the Hutton Field comprises a six-leg square platform in which four of the legs are anchor legs; wells are drilled through a template, which is eccentric in relation to the vertical axis of the platform; the production modules are located at the surface (see Fig.A.18.8).

TLP applications

The applications of TLP's are illustrated in Figs.A.18.3-A.18.6. TLP's will be seen to be of particular interest where:

- the number of wells is around 30-40;
- water depths exceed 200 m;
- payloads are around 10 000 to 30 000 tonnes.

Chapter 7

Product Dispatch

OFFSHORE LOADING STATIONS AND DESPATCH SYSTEMS: STATE-OF-THE-ART

Offshore dispatch and loading stations concern different products and different transfer and mooring conditions. These systems are practically the same, irrespective of whether designed for hydrocarbon field production, or processed product transfer (in some cases, such processing can be carried out offshore).

Transfer Conditions

This review is not intended to take into account port transfer type conditions where the take-off vessel is moored either at a wharf or jetty, or at a buoy. For sea mooring purposes, it has been necessary to design single-point mooring systems (SPM), allowing the vessel to turn and thereby reducing the stresses caused by the action of the elements (waves, winds, currents, etc.). Two principal cases will therefore be considered with respect to the take-off vessel mooring conditions.

Single-point mooring (SPM)*

The main components of this system, irrespective of the location of storage (offshore or onshore), are:

(a) mooring:
- an anchored buoy or spar buoy
- a mooring line for the take-off vessel on the buoy or spar

(b) transfer of fluids:
- a rotating joint (which may be single or multi-port) integrated in the SPM
- a riser connecting the subsea pipeline to the rotating joint supported by the buoy or spar
- a connecting hose between the rotating joint and take-off vessel.

Mooring to a Floating Storage Facility

In this case, the floating storage facility is a vessel, itself permanently secured to a single-point mooring system. (This type of system is often known as SBS or single buoy storage.)

For the purposes of this review, we shall consider the storage facility anchorage and the transfer of products from storage to the take-off vessel. The floating storage facility may also serve as a carrier for the product processing plant. (This type of system is known either as FPSO - Floating Production Storage Offloading or SPT - Storage Production Terminal.)

---

* See Appendixes 19 and 20, in which the subject of dispatch systems is further developed.

Our review will not, however, consider the transfer of production to the storage tanker. The take-off tanker is then moored alongside to the storage tanker either close alongside, or in tandem. The fluid may be transferred between the storage tanker and the take-off tanker either by floating hose, or by aerial hose.

The Products Transferred

The nature of the products is such as to impose temperature and pressure conditions on the transfer operation.

(1) Liquid products at ambient temperature and low pressure. This applies in general to degassed crude oils, at customary selling conditions.
(2) Liquefied petroleum gases (LPGs), where storage and transport are carried out at low pressure but at temperatures of around -50°C.
(3) Liquefied natural gas (LNG) for which there is as yet no loading station in existence of the kind defined above. Current design work on the construction of barge-mounted LNG plants pose the problem of construction of appropriate loading stations.

Hydrocarbons may be dispatched by a number of techniques, for which risk analysis is desirable.

Safety and Reliability

For the simplest systems, the drawing of fault trees should provide a risk analysis leading to a rational approach to the notions of safety and reliability. There are two possible types of risk analysis:

(1) Qualitative analysis (e.g., of pollution risk) with the designed aim to indicate the combination of circumstances potentially giving rise to such risk;
(2) Quantitative analysis (e.g., of the risk of disconnection of the floating support) as an aid to decision-making on whether or not to mobilise an assistance tug or keep a propulsion system on board a permanently anchored vessel (in the case of an SBS or FPSO).

THE SAFETY AND RELIABILITY OF OFFSHORE FLUID TRANSFER LINES

A Brief Description of the Main Types of System

There are three main types of system: CALM; SALM; and Compliant tower.

CALM (catenary anchor leg mooring)

This involves a buoy with a fixed rotating part anchored to the sea-floor by chains. Flexible hoses lead from the sea-floor to this part of the buoy, their configuration allowing the buoy to move. A rotary table is provided on the fixed rotating part. The mooring lines for the tanker and the floating flexible hoses to the tanker are taken from this rotary table.

SALM (single anchor leg mooring)

In this case, the buoy, anchored to the sea-floor by a single connection, is used solely for tanker mooring. The hoses are connected to a subsea rotating joint and rise to the surface without passing through the buoy, floating for a certain distance and then connecting to the tanker.

Compliant Tower

In extremely difficult environments, a tower secured to the sea-floor by means of a universal joint is used for mooring the tanker and protecting the fluid risers from the sea-floor to the surface. The top of the tower has a rotary table to which the tanker is moored by means of mooring lines. This head supports a boom from which a hose connecting to the tanker is suspended.

Faults and Their Consequences

Faults experienced in use include pipe leaks, either in the pipe itself or at connecting joints, and breakage of the actual pipe structure. These faults occur in the event of pollution and involve an interruption in loading, the replacement of the defective pipe or pipes, and where necessary, recovery of the lost crude.

Other defects are encountered on routine inspection operations or major maintenance operations in the workshops, the most frequent being the following:

- lining blister;
- leaks at end connectors;
- kinking;
- tearing of the outer casing;
- cuts in the lining;
- damage to float gear;
- wear of reinforcing layers.

The consequences of these defects are either repair or replacement of the lines in question.

By way of an indication, in 1983, the average life of these lines was around 10-12 months for parts subject to the greatest stress, and 2-3 years for others. These results were obtained on extremely well maintained installations following substantial improvements over the past five years. The general consequences of these faults are, of course, of an economic nature:

- pipe replacement costs;
- plant immobilisation;
- the need to hold major stocks of spares;
- preventive maintenance and inspection costs.

Cause of Faults

The various causes of the faults are as follows:

- product unsuitable for its intended use;
- faulty manufacture;
- inappropriate handling and storage prior to use;
- faulty configuration (design or installation error);
- pipe at end of its service life;
- accidental damage.

Product Unsuitable for its Intended Use

This relates either to structures which are insufficiently strong, or materials, particularly in the lining, unsuitable for the fluid carried.

Faulty Manufacture

The most common faults are attributable to poor cohesion of the various layers in standard lengths and at end connectors.

Inappropriate Handling and Storage Prior to Use

In addition to the problems of external aggression (tearing), the most frequent problem is failure to comply with the minimum radius of curvature in handling with unsuitable systems.

Faulty Configuration

This problem is similar to that above, and manifests itself in failure to observe the specified radius of curvature for the pipe. It may be due to erroneous design or to installation which is not in accordance with the specifications.

Pipe at the End of its Service Life

This relates either to the fatigue limit of reinforcing materials, or ageing of the lining material in the presence of the fluid medium.

Accidental Damage

This relates to impact or wear between the pipes, against the support vessels, against the tanker, against the lines, etc.

Solutions to Date

(1) Improvements in technical specifications and quality control procedures have enabled supplies unsuitable for their intended use to be eliminated, only the best products currently available being used.
(2) Precise procedures for handling and storage enable damage associated with these causes to be avoided.
(3) Cumulative experience in the design of loading systems has reduced the risks associated with configuration design.
(4) Finally, preventive maintenance and inspection procedures enable signs of faults to be identified, in most cases before destruction is total.

In the best supervised systems, a daily visual check is carried out on floating lines, with a weekly check on subsea lines and a monthly check on subsea line configurations. In addition, systems are dismantled annually for onshore inspection, and subsea lines are systematically replaced every two years.

The programme described above has allowed the average service life of floating lines to be multiplied three or four times over a period of five years, with the exception of the end runs, where service life has been doubled on the tanker-side connection and has remained virtually unchanged on the buoy-side connection (in CALM systems).

Problems Still Outstanding

In spite of the various measures described above, the major causes of failure continuing to affect properly used products are pipe kinking, and lining blister or damage.

From work done, it would seem that short hoses (35 ft) have a shorter service life than long hoses (40 ft). In addition, large diameter hoses (20 in.) seem to last longer than smaller hoses (16 in.).

Without going into greater detail, it would appear that the discontinuities in stiffness associated with the connectors, and the inherent stiffness of the product play a significant part in its service life.

Conclusion

Safety associated with offshore transfer lines can be suitably assured by a programme comprising:

- precise specifications;
- suitable quality control;
- suitable handling procedures and installations;
- a comprehensive preventive maintenance system.

The above programme is, of course, likely to be applied at major terminals. But the resources required may not be available at the smaller terminals, thereby involving safety risks, with particular reference to pollution.

As far as reliability is concerned, improvements clearly need to be made - particularly to the end pipes (tanker and buoy side) which still exhibit a relatively short service life, and efforts should be directed towards improving resistance to kinking and lining blister. It would also be useful to study and understand the influence of stiffness and length between connectors on the service life of the products. These improvements should undoubtedly bring about a considerable reduction in terminal operating and maintenance costs.

TANKER BERTHING SAFETY

Many berthing accidents have occurred in recent years as a result of inadequate safety precautions. Pilots are thus requesting berthing aids to enable them to judge their approach speeds and distance for these operations. There are many aids already in existence, indicating the distance of the ship and the speed of lateral approach towards the berthing line. These use underwater sonar signals, aerial ultrasonic signals, and underwater radar transmission.

Practical experience has confirmed the opinion of many pilots, reluctant to place full confidence in sonar and ultrasonic measurements. The turbulence produced by the ship's propellers and the uncontrollable parasitic reflections cause loss of sonar signals and result in poor quality measurements.

The ultrasonic system has improved the situation by reducing interference in signal transmission. A third system has been conceived by Syminex, who have developed the Radar Berthing Aid System (RBAS) using two radars to measure the fore and aft distances of the vessel (by frequency modulation) and its speed (by Doppler effect).

The radar signal indication is transmitted directly by the hull of the approaching vessel, and the information is given to the pilot on the three colour radar screens making up the distance/speed combination and immediately indicating the speed to be correct (green), too fast (yellow) or dangerous (red).

Table 4 Berthing aid systems compared

	Syminex radar	Sonar	Ultrasonic
All equipment is above water and is easily accessible for maintenance and inspection	YES	NO	YES
Frequency of signals used for measurement transmission	13-14 GHz	200 KHz	10 KHz
Continuous transmission for distance measurement	YES, by frequency modulation	NO (pulse)	NO (pulse)
Continuous direct measurement of speed	YES, by Doppler	NO (deduced from change of distance with time)	NO (deduced from change of distance with time)
Interference by underwater noise, ships' propellers wash, turbulence and tidal currents	NO	YES	NO
Interference by ultrasonic noise (machines, helicopters etc.)	NO	NO	YES
Propagation speed change errors	NO	YES (temp. gradient in water and silt)	YES (temp. and humidity) i.e. rain, mist, fog and snow
Interference from other sources	NO (12-14 CHz KU frequency band is interference free at terminal sites with power levels used	YES (reflection from unwanted underwater objects)	YES (from birds flying near sensors mooring hawse and work boats)
Environmental effects	NO	YES (marine growth on underwater sensors requiring regular cleaning)	YES (effects sand and mineral dust conditions)
Range	Can be extended to 2000 metres	Short range	Limited range

Table 4 (continued)

Echo signal	Good for all types of vessels in all conditions due to excellent beam definition of aerials and special tracking circuit	Poor quality echoes from lowdraft lightly-laden vessels	Echo quality varies with type of vessel
Distance measurement accuracy	± 5 cm	Depends on propagation speed	Depends on propagation speed
Velocity measurement accuracy	± 0.5 cm/s	Worse than distance accuracy	Worse than distance accuracy
SEA-WATER 35% sal. / AIR			
Propagation speed 0°C 1449 m/s 331		58 m/s	10 m/s
Change with temp. 15°C 1507 341	NIL	39 m/s	8 m/s
30°C 1546 349			
TOTAL	NIL	97 m/s (or 6%)	18 m/s (or 5%)
Self calibrating	YES (automatic measuring cycle)	NO	NO
Underwater refraction of sonar signals due to temperature gradients resulting in complete loss of signals at critical angles	-	YES	-
Above water refraction of signals due to temperature gradients (hull temp. effect) resulting in complete loss of signals at critical angles	NO	-	YES

Table 4 (continued)

Drift away detection (drift alarm)	Standard direct measurements unaffected by changing environmental conditions	Dependent upon distance measurement-accuracy related to speed of riser at sonar signal propagation (water temp. changes and turbulence)	Dependent upon distance measurement-accuracy related to speed of sonic change with time (day and night)

This system has been developed on the basis of the all-weather landing systems used for the past 10 years by more than 20 airlines. Its reliability is proven with only minor routine maintenance being required. This new system of radar berthing aid for tankers was installed some months ago at the Fos oil terminal near Marseilles, and Syminex recently equipped the German port of Wilhelmshaven with this system.

SEA-LINES

Inspection, Maintenance and Fault and Leak Detection*

Pipeline inspection involves checking the entire length of the pipeline, possibly dynamically. The inspection can be external (aerial, underwater or surface) or internal (carried out with instrument pigs). Techniques are developing very quickly in inspection. Each particular operation is perfected and uses the most sophisticated means.

Inspection is to form an integral part of the construction and service use of structures. It allows maximum limitation of the risks of failure by preventive treatment, and the identification of zones requiring curative treatment. It therefore helps to ensure that production is carried out under the best possible conditions, checking the structure for satisfactory behaviour. This behaviour may be altered by the following effects:

- internal or external corrosion;
- fatigue or cracking;
- leaks;
- poor condition of external coatings - epoxy, coal tar, concrete, etc.;
- erosion;
- clogging with deposits or build-up of rubbish;
- poor positioning of the line on the sea-floor.

Inspection is carried out by specialist service companies, using procedures that comply with current laws and regulations, to ensure maximum safety for personnel and structures. It is performed under the control of official inspection authorities which certify approval of the structures.

Inspection may be classified into two categories: routine or systematic inspection, and inspection associated with a specific incident on the line (leak, operating fault, etc.). The first category covers:

----

* See also Appendix 21 for a further development of this subject.

- inspection of the cathodic protection;
- corrosion check;
- coating check (epoxy, concrete, etc.);
- leak detection;
- geometrical inspection of the line;
- check on the mechanical condition of the pipe;
- check on the position of the line on the floor;
- pipe burial check;
- cleanliness or pressure drop check.

These various checks may be carried out externally or internally.

Inspection associated with accidents involves the use of specific inspection facilities in line with the faults in question: i.e. leakage, buckling, etc.

Inspection Methods

In endeavouring to list the various inspection methods, we should begin by considering external inspection, which involves greater or lesser aerial, surface and/or underwater facilities, in all cases related to the conditions of approach (depth, location, weather conditions, etc.), followed by internal inspection which makes use of the natural path of the pipeline and is carried out by instrument pigs, and, as a result, is much easier.

Before engaging in any inspection exercise, it is essential to ensure that the problem has been correctly defined. This may seem obvious, but it greatly affects the success of any operation and cost optimisation. The inspection method best suited to the problem must then be identified, anticipating the secondary effects, to limit resource mobilisation efforts.

Finally, the company used should be the one offering the best conditions of performance and guarantees of success for the operation.

External Inspection Methods

Surface methods. A surface vessel is used to tow the 'fish' responsible for performing the inspection. Alternatively, it may carry out the inspection itself. The principles used with this method enable the following measurements to be taken:

- acoustic measurements (propagation of pressure waves under water) for monitoring pipes that are not deeply embedded, and for determining the profile of trenches and depths of burial;
- visual (photographic camera);
- measurement of the cathodic protection;
- leak detection by water sampling;
- magnetometry or measurement of variations in the magnetic field of the ground, for locating pipelines.

Underwater methods. These are carried from a nautical support facility, using the following resources:

- divers operating from the surface, possibly with independent propulsion systems;
- divers operating from a manned submersible;
- manned submersibles;
- remote controlled devices.

They are designed to perform the following operations:

- pipe location and tracking;
- visual and video or photographic inspection;

- identification of trench profiles and sections;
- identification of the depth of burial, burial heights and lengths;
- measurement of cathodic protection;
- leak detection and location;
- measurement of pipe thicknesses;
- check on the condition of coatings;
- check on corrosion and external mechanical condition.

Depending on the type of inspection required and, of course, the depth, the resources required may be divers or submersibles. The general rule is that 0-50 m involves divers and 50-300 m involves submersibles, manned or otherwise. Manned submersibles may be considered for working to depths of 1000 m. Beyond this depth, remote controlled systems must be used.

These services and the equipment required for carrying out this type of inspection are available from a few European service companies, but mainly from American companies. They provide information on the condition of the pipelines. Since the ultimate aim of inspection is to survey general soundness of the item with a view to identifying any action required, the current state of the art still leaves a number of areas of weakness, which research and development efforts are endeavouring to reduce.

It would, for example, be useful if European companies could offer the full package of integrated methods of line inspection using the same carrying facility, with a view to limiting costs. It would also be useful to develop projects with a view to standardising criteria for certifying the leak and pressure tightness of pipelines using computerised testing programmes.

It is also suggested that methods aimed at improving the quality and protection of burial trenches be developed. The cost of the resources required for offshore or subsea techniques is, however, a major obstacle to the advancement of projects in this field. Subsea repair work, very similar to inspection, with respect to the resources used, should also experience rapid development to provide a timely response to the major line repair problems already being encountered.

In this same connection, mention might be made of:

(1) potential improvements to materials (with particular reference to heat and cold insulation, stress and corrosion resistance);
(2) the development of subsea connection systems (extremely important with respect to underwater repairs and deep-water pipelines);
(3) the creation of a method of reducing the weight of pipes for pick-up and repair, and in the long term, using robots for welding or pipe connection repair;
(4) the continuation of biological studies into deep-water diving, already started by COMEX.

Internal Methods

Inspection is carried out by instrument pigs. These pigs are propelled by the fluid (liquid or gas) and generally comprise the following components:

- a propulsion unit - this is the part which prevents any leaks by means of collars, and is subject to pressure, driving the entire unit.
- a measuring unit - this is the sensitive instrument portion designed to carry out the actual measurements.
- a power pack, generally in the form of batteries, to supply the instruments.

Types of inspection suitable for performance in this way are listed below.

(1) Geometrical pipeline check (designed to ascertain the extent and location of any distortion suffered by the pipeline).
(2) Corrosion inspection (corrosion is the principal enemy of steel pipes and inspection activities designed to detect and assess corrosion are important. They are carried out by instrument pigs using the principle of magnetic fields, to determine the residual thickness of the metal on the walls).
(3) Check for mechanical defects (cracks, adhesion, etc.).
(4) Check on cathodic protection or the presence of anodes (this process is currently under development, and is not yet operational).
(5) Check on location (three axis positional survey) - not yet operational.
(6) Measurement of the internal temperature of pipes (especially for gas lines, to detect and prevent hydrate formation) - not yet operational.
(7) Check for the presence of water - not yet operational.
(8) Leak check.

As will be seen from this summary list, a number of methods of internal inspection by instrument pig are currently under development: these methods have the advantage of being relatively easy to use, the resources employed being less substantial and hence less costly. Unfortunately, in this field, the market is virtually monopolised by the Americans. The only types of pig available in Europe are sizing pigs. There are no French companies manufacturing ordinary pigs. Investigations cover a relatively broad area, as it is quite conceivable for a number of operations to be performed from inside pipelines, from inspection through to repair, as a means of providing a low-cost survey of the general condition of the lines and identifying the internal facilities required to remedy any defects.

It would be useful to have a network in Europe with a facility for simulating or reproducing the conditions of use of instrument pigs on a small scale, checking their suitability for the various circumstances encountered, and training operating personnel in their use. This would enable the rate of progress on design projects to be greatly accelerated through joint efforts and experimentation. We hope, therefore, to see the development of instrument pig systems to cope with:

(1) the chemical treatment of pipelines for leak detection, location and possibly repair;
(2) pipes to be calibrated and re-shaped;
(3) tight internal plugs to be made at selected locations, to isolate particular sections of the line.

In addition, instrument pig development efforts could be directed towards improvements in the metering of distances covered and the recording of information or measurements and their external transmission, these improvements resulting in a reliable, obedient, competent and reusable shuttle system.

The various methods used for leak detection are listed below.

Aerial methods. Aircraft or helicopters can only detect relatively major slicks. They can, however, be quickly deployed, and can be used to identify more restricted areas for investigation. The methods used are either aerial photography, based on the contrast between the oil slicks and the sea-water, or a thermographic or infra-red method.

Surface methods. These involve the use of a boat towing a detector device and sampling the water for subsequent analysis to detect any traces of hydrocarbons, or alternatively using a sonar system to detect bubbles or suspended particles.

Underwater methods. Visual inspection, where the zone is demarcated by divers or piloted submersibles.

(a) Acoustic methods: using the system of the pressure drop at the leak, or the detection of corresponding frequencies.
(b) Tracer methods: a tracer is injected into the line and should normally be partly discharged at the leak. The tracer may be either a dyestuff (e.g., fluorescin) or a short life radioactive product.

Internal methods.

(a) Ultrasonic method: the instrument pig records noise and distinguishes between the normal 'signal' noise and the noise of a leak across its path.
(b) Detection by short life radioactive tracers: the instrument pig detects the traces at the leak.
(c) Pressure drop method: a pressure tight pig provides an artificial separation between two sections of the line. The location and extent of the leak is then established by successive approximation, measuring the pressures at each end.

Conclusions

There are numerous methods of inspecting lines. Hence, the most suitable methodology can be selected for each particular case. However, to provide a complete range of operational methods, many additional methods still require development. It is important for owners to be fully aware of the condition of their pipelines at all times and, consequently, to decide which preventive inspection methods will provide them with the best information, to enable them to anticipate developments, prevent accidents and operate their installations in the best possible manner.

Pipeline Repairs - Techniques

General

Irrespective of whether structures are prefabricated, surface or submerged type, and particularly pipelines, they will always require work to be carried out underwater, for initial installation, inspection or repair purposes. Where pipelines require repair, the main causes are external and internal corrosion and fatigue, due to severe service conditions, impact, and various accidents - during laying or burial, anchor impact, etc. When damage is identified, the following action should be taken.

(1) A preliminary inspection to be carried out with the aid of remotely operated devices and divers to locate the damaged part and determine the following:

(a) nature of the floor, with a view to defining any uncovering methods required;
(b) environmental conditions: depth, state of sea, distance, with a view to defining the type of support vessel and equipment required;
(c) investigation of the damage;
(d) ascertainment of the length of pipe to be replaced.

(2) From this information, an engineering study should be undertaken with the aim of selecting the tie-in technique, and optimising repair time and costs.

The following aspects are to be covered: (1) procedures for recovering the damaged section; (2) design of the new part (length and handling procedures); (3) alignment procedures (positioning and moving underwater lifting gear); (4) definition of the operating procedures; and (5) definition of the equipment required.

Environmental Conditions

Connection work on pipelines falls into two categories.

(1) Mid-line tie-in consists of making or re-making the continuity of a line where the latter has been broken, for example as a result of one of the instances of damage specified above. The dimensions of pipelines in this configuration are generally 6 in. to 36 in. in diameter, with thicknesses of 8-50 mm. Steels are types API 5L GRB to X 65.

(2) Riser connection consists of making a continuous line running down the length of a jacket leg. This elbowed part of the line is termed the spool piece. The dimensions of pipes in this configuration are generally 6 in. to 36 in. in diameter, with thicknesses of 10-50 mm. The steels used are types X 42 to X 65. Working depths may be to 200 m (STATPIPE Deep Dive Development Project).

Connection Techniques

(a) Bolted flange assembly or mechanical connection. This is the oldest method. The ends to be connected are fitted at the surface with welded flanges and are bolted in the water either by scuba divers or by pre-saturated divers, depending on the depth. These divers use hydraulic impact spanners powered from the surface to achieve the necessary tightening force. This type of flanged assembly may, however, lose its ability to prevent leaks after a few years, under the combined effects of corrosion and mechanical stress, causing leaks and requiring repair.

This technique of step-by-step mechanical assembly could be a relatively easy subject for automation or remote control. This could be of particular interest for deep-sea projects.

(b) Atmospheric welding or, to be more precise, underwater welding in a dry atmosphere at atmospheric pressure. The electrodes are transferred in a gaseous atmosphere of air at atmospheric pressure. Welding conditions are thus the same as at the surface. This method (the WELDAP process) is said to enable welding work to be carried out at depths exceeding 300 m. An experimental weld was produced under these conditions in 1977 on a sleeve, diameter 20 in. and thickness 20 mm, at a depth of 300 m. This technology will only really become usable following a development programme aimed at improving leaktight conditions hence manpower safety, and reducing costs, which still remain excessively high.

(c) Hyperbaric welding. This technique has been operational for more than ten years and is widely used by oil companies in view of its reliability and durability. These developments and the results obtained from welding under hyperbaric conditions allow joints to be produced of a quality equivalent to those welded at the surface, at atmospheric pressure. Different, highly specialised facilities are used according to the particular application. Men carrying out this work must be divers/pipe fitters/welders/inspectors. They are classified in the highly skilled category.

Electrode transfer is carried out in a gaseous atmosphere other than air, and at a pressure corresponding to the pressure of the subsea environment in which the connection is to be made. It is the only method of subsea welding, and is described in Appendix 23.

Part III

Inspection - Maintenance - Repair Technology (IMR)

Chapter 8

The Sounding and Maintenance of Concrete and Steel Structures

In 1982, offshore oil activities involved some 4000 platforms with thousands of kilometres of pipelines scouring the sea-floors. Over the years, these various structures suffer mechanical and chemical stresses under their own loads and a frequently hostile environment. The 70s, with the major development of fields in the North Sea and the start of production, first saw the need, then the necessity, of regular inspections on structures exposed to the combined actions of marine corrosion, wave stress cycles and external mechanical stresses caused by the wind and freak waves in heavy storms.

Inspection programmes applying non-destructive testing techniques are now standard and are bound to develop further in future years. Inspection is justified by two motives (irrespective of the improvement in safety for site personnel):

(1) the economic motive; the aim being to prevent any accident liable to cause a production shutdown or loss of operation;

(2) the regulations motive in areas where government authorities have imposed stringent regulations, or to meet conditions required by insurance companies, certifying bodies or even the owners of the installations themselves.

The integrity of a structure after construction is based on three types of action: (1) regular maintenance; (2) inspection, which may involve specific maintenance activities; and (3) repairs.

INSPECTION WORK*

The initial appearance of defects and the further development of any defects or damage liable to affect the life of an underwater structure must be detected as early as possible to enable work to be carried out quickly and efficiently, where required. This is the object of inspection programmes.

What are the characteristic defects for which structures should be inspected? They include:

- mechanical impact occurring in ship loading or offloading operations;
- marine vegetation (extent and nature), increasing tensile loads;
- fatigue cracking in nodes subject to major stress;
- local corrosion due to poor cathodic protection;
- anode condition;
- preferential corrosion in zones heavily affected on welding;
- wear caused by abrasion and scouring;
- ferrous debris coming into contact with the structure, being a cause of corrosion by electrolysis.

---

* For the inspection of concrete structures see Appendix 22 for a summary of the most recent data.

The techniques used in subsea inspection are mainly onshore techniques adapted to the subsea hyperbaric environment (see Tables 5 and 6). Marine fouling and cover by various organisms: the structure needs to be cleaned to allow inspectors to do their work properly. The most widespread inspection technique is visual inspection. This enables the general condition of the structure to be assessed and the major defects to be detected - 35 mm and 70 mm photographic films are processed on the surface. Black and white and colour video enables all or part of the results to be kept.

Table 5.

Methods	Object	Results	Comments
Measurement of potential	Check on cathodic protection.	Potential reading.	In situ or surface reading
Photography 35 mm 70 mm	Visual perception of the results of inspection.	Photograph, with possibility of enlargement, making up photographic mosaics.	Macro photographs possible in both zones. Developed on site. Better definition than 35 mm.
Video system black and white colour	Continuous record of inspection.	Recorded cassettes allowing inspection to be carried out by specialist personnel.	Good definition. Better definition than black and white.
Ultrasonic thickness measurement	Check on metal thickness.	Digital reading in situ or on the surface.	Prior adjustment required. Choice of sensor important.

Inspection may alternatively be carried out by remote-controlled systems. This field of activity may be assimilated to visual inspection. Where defects are too fine or inside the material, they cannot be detected by the eye. Here, special equipment has to be used. The operator then interprets the result of a physical phenomenon:

- potential current measurement;
- thickness measurement by ultrasonic inspection;
- magnetic particle inspection with the fault reading under ultraviolet light with the subsequent possibility, however, of visual indication using special ink under white light;
- temperature measurement.

Checking for faults inside the material involves the use of more elaborate techniques: ultrasonic fault detection, and gamma ray radiography, Iridium 192, etc.

Table 6.

Methods	Object	Results	Comments
Ultrasonic fault detection	Detection, location and assessment of internal faults.	Major faults located.	Satisfactory results achieved by surface monitoring repeater with good quality of communication.
Magnetic particle inspection	Detection of surface cracks.	Indication of cracks in or emerging at the surface, visible by UV rays.	Possibility of photographic record using special ink. Faults less than approx. 2 mm from the surface.
Temperature measurement	Detection of high or low temperature points.	-	-
Gamma ray radiography Ir 192	Inspection of connecting welds. Examination of walls/pipes.	Permanent inspection document.	Interpretation by highly qualified personnel/dangers due to ionising radiation.

At the end of inspection a final report is produced. This is prepared from the inspection documents produced in the preparatory phase.

Sounding of Welded Joints on Metal Platforms

Many years of highly systematic research and development work on the use of acoustic emission as a means of monitoring the welded joints of metal structures, and detecting and monitoring the development of any cracks which may occur in the weld seams of such joints, have led to the following conclusions.

(1) The marine and industrial environment is not a technical obstacle to the use of acoustic emission (tests carried out on ASHTARI in 1979).
(2) The acoustic emission technique of sounding is industrially and economically feasible.
(3) It is possible, thanks to the data processing work carried out in 1981, to detect a 64 mm crack in a welded joint on an offshore platform (tests on a platform in Field L7).
(4) Acoustic emission monitoring of a localised crack is both possible and economically feasible.
(5) *A priori* surveillance of certain joints in metal platforms is possible, economically viable and compatible with the inspection philosophy involving detailed surveillance of zones (joints) subject to the greatest stresses.

Although the above points are undeniably positive, this process is unlikely to become more widely used under current conditions, for the following reasons.

(1) None of the equipment used so far would be used in the operational system (essentially this has been laboratory equipment, unsuitable for site use and with limited performance characteristics).
(2) The platform test (on L7, 1981) was carried out under favourable conditions (with sensors excessively close to the crack being detected - the joint was totally underwater, and not in the tidal range subject to greater noise and exposure to wave stresses).
(3) Certain components designed or developed in 1981 have not yet undergone field tests (transmitter-receiver sensors, magnetic suction cup sensor fitting systems, electronic acoustic emission data recording and automatic system monitoring modules).

Thus although the basis is excellent and acknowledged as such, the potential of the acoustic emission technique cannot yet be used without technical and financial risk. Reliability tests are therefore needed in the form of monitored sound emission tests in the field.

REPAIR WORK

In certain cases, the results of inspection will be such as to require remedial action to be taken. A repair formula is then to be proposed, taking into account the cause of the damage. We have already identified the principal causes of damage in the list of characteristic defects for which structures are inspected. These were: corrosion (undoubtedly the most important and liable to vary considerably according to the depth and nature of stress); fatigue (most frequently due to underestimation of stress in the area of welded joints); and finally, impact and accidents.

There are two main types of repair techniques used: mechanical repair and repair by welding.

Mechanical Repair

These consist, for example, of bolting shells onto damaged struts to seal and reinforce the structure. This type of repair is often regarded as temporary, but is quick and relatively cheap to perform.

Other techniques are currently under investigation - in particular a system of repair by the placement of shells on the structure, and then grouting with concrete.

Subsea Welding Repairs

There are two possibilities here:

(1) In-water welding, using stick electrodes coated with varnish. The weld quality is still poor, in spite of the major improvements made over the past two years.

(2) Dry wèlding in a hyperbaric chamber at a pressure equal to ambient pressure, or at atmospheric pressure. The quality of weld is excellent, conforming to the best standards required by inspection authorities (see Appendix 23).

Particular mention should be made at this point of repairs to structures in concrete. This type of work essentially uses resins designed to provide continuous protection against corrosion for reinforcements or tendons.

MAINTENANCE DATA BANKS

The inspection and repair information on a given structure forms a mass of data requiring careful filing and review throughout the life of the structure. In recent years, companies have therefore set up data banks in which to store this information. These may be defined as an aid to maintenance.

The primary function of data banks is to commit structures to memory. The complexity of structures and work performed results in the storage of a considerable quantity of information. This information will provide an understanding of the structure not only in space, but also in time. Access to the memory is assisted by graphics, and involves homing on particular points, providing for rapid familiarisation with the structure and retrieval of information relating to a particular item. Photographic and video references are also kept.

Once the results of work performed have been inspected in the system, an inspection report can be produced. The data banks also allow planning to be carried out on the basis of specifications, thus providing a facility for monitoring work, in real time.

FORESEEABLE TECHNOLOGICAL DEVELOPMENTS

In this area, the main basic techniques are already established. Hence foreseeable developments will relate to a further mastery of these various technologies, possibly developing in two main directions, namely:

(1) greater efficiency and improved productivity, in particular via progressive automation and increasingly widespread use of cable type remote controlled systems to carry a range of inspection and monitoring sensors;
(2) perfection of the basic techniques mentioned above, by seeking greater reliability of instruments, improving performance and operational procedures, and computerising information processing systems.

Chapter 9

Subsea Installation Inspection Techniques

REVIEW OF PROCEDURES USED

General

The capital invested in drilling and development structures and oil and gas transport systems offshore currently amounts to several billion dollars. Massive metal or concrete structures of a size inconceivable a few years ago have been developed to extract petroleum resources from the bottom of the sea, not only under the best economic conditions, but also under the best conditions from an environmental protection viewpoint.

Developing reserves means initially designing, building and installing platforms to be used for drilling hydrocarbon production wells and installing the equipment required for oil/gas separation. It also requires provision to be made for despatching the product either by laying a subsea pipeline, or by installing a floating loading station.

These offshore facilities are subject to various forms of deterioration caused by problems of fatigue (wave resistance), erosion, corrosion, impact with other floating items, etc. Whether for economic reasons (avoiding production shutdowns), or for safety or environmental protection, inspection of the condition of these facilities and repairs to damaged parts will, in future years, constitute a major proportion of the work carried out by diving companies.

Offshore structures are already subject to extremely strict regulations (Mineral working, Offshore installation, Act 1974) requiring each structure to have a certificate of worthiness covering both their initial design and the monitoring of damage caused by the marine environment or their particular conditions of use (e.g., corrosion of pipelines by the oil transported).

The inspection of offshore installations uses all possible techniques to provide a precise survey of the condition of offshore or subsea structures in wood, metal or concrete. In this area, the work that can be performed by divers is still developing, but the future of remote-controlled systems such as manned submersibles seems promising, particularly for deep-sea applications.

Particular mention should be made here of the primary importance of the reliability of information provided by 'subsea inspectors': this property is a function of the way in which measurements are taken and interpreted, requiring an adequate knowledge of marine corrosion problems and non-destructive testing techniques.

Corrosion of Offshore Structures

Sea water is a particularly favourable environment for corrosion phenomena.

(a) It is a living medium, containing varied organic compounds in both the animal and the vegetable realms. Naturally settling on any underwater carrier, they can affect corrosion phenomena either directly (change in pH) or indirectly (differential aeration).
(b) It contains substantial quantities of dissolved salts (30-35 g/l) and its conductivity is approximately 50 times that of fresh water.
(c) The dissolved gases, particularly oxygen (5-8 cm^3/l between 0 and 25°C), play an important role in the corrosion process.

All of the above parameters vary as a function of depth.

In time, the majority of materials react with their environment, changing in structure. In the case of metals this occurs through corrosion, and in the case of concrete, polymers, etc. through weathering.

In metals, corrosion (whether chemical or electro-chemical) can have three consequences:

(1) elimination of the metal (e.g., Zn);
(2) formation of an oxide (e.g., Fe_2O_3);
(3) formation of another compound (verdigris).

In general, corrosion tends to return the metal to the form of its ore. In offshore structures, corrosion will differ substantially according to whether the part in question is in the superstructure (saline mists), in a zone subject to agitation (the tidal or splash zone, rudder or propeller zone etc.), or at depth.

In the splash zone, corrosion is at its maximum, primarily because of permanent contact with a highly aerated sea-water, combined with erosion due to the salt water spray. Corrosion is also substantial in the mud zone, where a new type of concentration pile is appearing, due to the sulphur compounds.

In the water, fatigue phenomena prevail, and inspection should be concentrated primarily on the welds. In concrete structures, the rate of deterioration is related to the cohesion of the concrete around the reinforcements. This deterioration may be chemical or mechanical. Wood is considerably more sensitive to erosion, but also to attack by microorganisms (i.e. ship worms).

Subsea Inspection Equipment

Major work has already been undertaken with a view to developing, modifying or adapting non-destructive testing equipment and techniques. Ultrasonic thickness gauging, magnetic particle weld inspection, and all TV and photographic observation techniques are already in widespread use for the inspection of offshore structures. Although each technique is specific to a particular type of observation, in most cases several will be used to gain a better understanding of the part in question. A full inspection of an offshore structure may include:

(a) a visual inspection to determine the general condition of the structure, any evident damage, the quantity of foulants, the state of the cathodic protection system;
(b) a detailed examination, after cleaning, of certain critical points (welds, joints, etc.) or points repaired in the past (in general, this is done by closed circuit television, a surface inspector remaining in communication with the diver);

(c) as a function of this inspection, more detailed information may be required to assess the presence or seriousness of certain defects;
(d) finally, potential measurements may be taken systematically, to measure the degree of protection of the structure.

A very precise specification is drawn up before each inspection is undertaken.

Surface Preparation

It is essential to prepare the surface to be examined before undertaking any non-destructive testing. The quality of a weld photograph, for example, will be greatly dependent on the care with which the weld was cleaned. There are several techniques of doing this.

Brushing or blasting. This may be manual or mechanical. For mechanical brushing and needle jet blasting, a hydraulic tool is used in conjunction with an underwater unit. An underwater hydraulic unit may be used to depths of up to 300 m; the power available at the tool is independent of the working depth, as a result of oil tank counterpressurising. Electricity is supplied from the surface through an electrical cabinet and a connecting cable.

Very high pressure water jet. This is the best process for preparing surfaces prior to inspection, as it cleans the surface thoroughly without attacking either the structure or the underlying metal. The system generally employed uses hoses with different nozzle fitments to suit the work in question. An equalising jet system allows it to be used underwater. The hose is supplied from the surface of an HP pump (120 l/min at 420 b), driven by a 140 HP Diesel motor. The working pressure is generally 315 bars (4500 psi) for cleaning off foulants, or 420 bars (6000 psi) for blasting welds. The addition of iron shot at the gun improves blasting efficiency.

Cathodic Protection Inspection

All offshore production and oil and gas transportation facilities have cathodic protection against the various corrosion risks affecting their underwater and buried parts.

Serious inspection of the electrical condition of platforms and sea lines and on the proper functioning of their cathodic protection systems is essential and, indeed, is frequently required by current regulations.

Measurement of the electrical potential either of the structure or of the anodes is carried out by placing a voltmeter in an electrical circuit connecting the mass of the structure protected (+ve pole) to a submerged reference electrode (-ve pole). The conditions of contact with the mass of structure must be good, to avoid stray contact resistance.

Many fixed installations have a disadvantage in giving only an overall measurement, and not the potential in the recesses (these are the critical points, since they are the least affected by the protective currents, and the most sensitive to cracking as a result of fatigue corrosion). Here, an electrode is used, carried by a diver, with either a surface voltmeter or the electrode and voltmeter combined in a single instrument. This is the bathycorrometer. This provides for potential measurement; the diver is fully independent and has no electrical link with the surface. The instrument can be easily held in one hand; the titanium contact electrode is held in the other. The reading is taken on a 0.2 V voltmeter with 0.05 V graduations (accuracy $\pm 2\%$). The housing is totally in pressure-resistant plastic.

Measurement of the protection of a pipe can be carried out along the entire pipe by following it with an observation submersible, or using a remote controlled device equipped with a track pipe, and recording the current density curves with the aid of a detector, specially installed on the submersible.

Corrosion Measurement

Corrosion may be manifested either in a reduction in thickness (general corrosion), or in the presence of defects (cracks, caverns, chinks, etc.). In the former instance, regular measurement of the thickness of structural members (generally by ultrasonic means) will enable the efficiency of its protection to be determined. In the latter instance, techniques which are more difficult to use (e.g., MPI, gamma ray radiography) can serve to reveal the presence of flaws.

Thickness gauging. A number of ultrasonic devices are currently available on the market. COMEX use the D-metre, a highly sensitive device contained in a pressure-resistant aluminium housing.

The instrument is held in a frame, with plastic systems connecting to the external adjustment controls allowing the instrument to be reset to zero and corrected for the type of material under examination. The probe is connected to the housing by a watertight cable.

Fault detection. When a fault is found in a structure (generally at a weld) it is essential to determine its extent. Various techniques are used here.

(1) Ultrasonic inspection. Ultrasonic fault detection with a display on a cathode screen is a delicate but extremely useful method, requiring highly competent divers. The re-reading of a video recording on the surface provides a facility for closer analysis of the inspection results.

(2) Magnetic particle inspection (MPI). This is a simple and efficient method of detecting faults (cracks), even before they have reached critical dimensions. From a tank on the surface, suspended fluorescent ferromagnetic particles are sent to the surface to be examined, between the poles of a permanent magnet. Any lack of homogeneity, such as blowholes, cracks or inclusions produces distortion of the induced magnetic field in a magnetic material. The fluorescent ferromagnetic powder will therefore deposit itself regularly over the magnetic field, except in the location of the fault: this will be revealed by lighting, using a black light source (close to UV).

The device used comprises a UV lamp in a metal housing, with a trigger type system on its handle for spraying the liquid containing the powder. This method is of particular interest for faults close to the surface. It can only be used by divers specially trained in the use of this technique.

A technique of photographing faults revealed by magnetic particle inspection has been developed (with faults showing red in a Xenon flash light).

(3) Radiography. The instruments used are of the Iridium 192 gamma emission type. The source is enclosed within a can of depleted uranium remotely actuated by compressed air or cable control. The detector returns the information to the surface for a read-out.

Subsea observation. Observation transmitted to the surface (by TV) or carried to the surface (by photograph) is of prime importance, enabling an inspector on the surface to decide on further action required pursuant to such inspection.

Visual inspection forms the basis of all subsea inspection. No measures could be conceived without observing the site at which they are to be undertaken.

Dimensional checking generally provides additional information (width of crack, depth of crevice, dimensions of sacrificial anodes).

(1) TV inspection. Inspection of structures by closed circuit television is widely used by means of divers or remote controlled systems. The majority of systems comprise:

(a) a portable underwater TV with a projector (TC 125 black and white camera, or JVC C4 8800 V colour camera in a resistant housing, with surface controls);
(b) a diver-surface communication system;
(c) an umbilical to supply power to the camera and to transmit messages and TV images;
(d) a cathode screen on the surface, with video-sound recording facilities, enabling the images observed to be exactly reproduced.

(2) Subsea photography. This is now a well-known technique, with all sorts of cameras available on the market, such as the Nikonos, or housings enabling the most sophisticated cameras to be used. Macro-photography enables interesting information to be collected on the condition of structures.

After cleaning and blasting, it is possible, (e.g., with the aid of a 'Hydroscan' camera) to provide a full picture of a weld, after carefully labelling it at regular intervals. This camera is loaded with 250 exposure automatically winding cassettes.

The adaptation of the most sophisticated methods such as photogrammetry will, in the near future, enable more complete, therefore more reliable, information to be collected.

The inspection of offshore structures will increasingly require the involvement of skilled personnel, on the one hand familiar with the use of frequently delicate equipment, and on the other able to interpret the results of inspection in situ.

Detailed training in each technique, with a good theoretical knowledge and many years of practical experience will be required to turn a good diver into an efficient subsea inspector.

FORESEEABLE DEVELOPMENTS

The development of offshore oil exploration has resulted in large numbers of offshore platforms whose safety is subject to regular and effective inspection. This inspection, particularly where underwater, raises numerous technical and economic problems; the two aspects are closely connected. Considerable work has thus been undertaken in recent years with a view to perfecting strategies and inspection tools, to provide an appropriate response to safety needs. Promising results have already been obtained. New methods have been successfully tested and then marketed. Using the most recent non-destructive testing (NDT) techniques, modern inspection strategies are now being studied by numerous oil companies. There is still a great deal to be done, but the paths for future development now seem clearly defined.

Having passed the stage of simple visual inspection by divers (this inspection technique rapidly proved inadequate due primarily to the marine fouling covering the structures and concealing defects, the cloudiness of the water and the reduction in human faculties underwater) subsea inspection companies have endeavoured to adapt laboratory NDT processes such as ultrasonic or magnetic particle inspection for offshore use. These endeavours have met with widely varying technical success: in practice, only magnetic particle inspection is used on an industrial scale for detecting underwater cracks. Techniques based on the measurement of eddy currents now appear to be competing with this method. Ultrasonic measurement is limited to thickness gauging, of definite interest in assessing the state of corrosion of metal items underwater (jacket members, risers, pipelines, etc.).

All of these techniques do however have a major disadvantage: they involve prior cleaning of zones to be inspected. This takes a considerable length of time and requires highly specialised divers, and is extremely expensive to perform. These techniques are therefore applied within the framework of limited inspection strategies: only around 2% of the welds in a structure, subject to the greatest stresses, are regularly subjected to magnetic particle inspection. Theory shows that with this procedure, the chances of failing to detect cracks (including major cracks) remain enormous. In recent years, theory has been largely confirmed by practice, and this form of inspection has been seriously called into question and its efficiency has proved inadequate.

Major efforts have, therefore, been made in various parts of the world, but particularly in Europe, to reconsider the problem and provide a more solid and realistic foundation. All developments in new inspection strategies and techniques are based on the following ideas.

(1) An inspection technique or techniques must be adapted to each individual type of marine structure. This produces the greatest efficiency at the lowest cost.

(2) Every new technique must be developed with a view to subsequent automation. The use of divers is an increasingly unacceptable constraint, except for shallow locations (less than 60 m).

This explains the appearance of a variety of NDT methods, each suited, sometimes exclusively, to particular types of structure. Thus vibro-detection has appeared for existing metal jackets, pressure detection for new metal jackets, acoustic emission for metal structures and shortly TLPs, vibrational examination on all platforms, including gravity structures, etc.

Vibro-detection

This is a non-destructive testing method for jacket platforms, developed between 1978 and 1980. Fast, and hence economical, it enables the entire platform to be checked in a matter of days, even hours. It will detect all significant cracks (i.e. all cracks which are dangerous for the structure), irrespective of location. Used industrially for the first time in October 1982 on the seven platforms in the Total-ABK Field (Abu Al Bu Koosh) in the Arabian Gulf, it enabled these structures to be exhaustively inspected over a period of 13 days site work. The principle of vibro-detection is very simple: the dynamic parameters of each member of the jacket are measured; these parameters are sensitive to the existence of local cracks and hence reveal such existence. It is quick because it needs no preliminary cleaning and it is also very simple to apply, requiring no specialist ability on the part of the divers, which also adds to its economic advantages. Above all, it allows the method to be automatically applied by a remotely controlled system.

Prospects for the development of vibro-detection are considerable. Miniaturisation and automation of equipment have already been undertaken. The next stage is the production of a vibro-detector operated by a remote controlled device. Its advantages are substantial - particularly bearing in mind that platforms will shortly be installed at 200 or 300 m depths.

Pressure Detection

This represents an alternative solution for inspecting metal jackets. The principle once again is very simple: each tubular member is inflated with air to a pressure exceeding that of the surrounding water. Therefore, if the wall becomes cracked, the member will deflate and a pressure sensor will release an ultrasonic alarm. Each sensor is independent: there is no cable connecting it to the surface unit. This avoids the majority of the maintenance problems associated with underwater instrumentation systems connected to the cable work.

The advantage of this technique is that it minimises the use of divers (for installation and maintenance), making it a highly promising system for deep-water structures. Since its use directly affects the sizing of the various members, it is limited to new jacket constructions.

Prototypes for the sensor and surface unit (for receiving and decoding ultrasonic alarm signals) have been successfully constructed and tested in a tank environment. The next stage is instrumentation on a platform. This is scheduled to take place in 1984.

Acoustic Emission

This is nowadays a relatively widely used technique, particularly in oil refineries. Its adaptation for offshore use has raised serious difficulties, its first commercial underwater application having taken place in the North Sea in 1982. This (the most direct offshore application of acoustic emission) involved monitoring the development of certain cracks detected by visual inspection, until such time as they could be repaired. To do this, a joint was fitted with a number of suitable detectors, connected by cable to a data acquisition unit on the surface. Subsequent processing of the signals collected in this way enabled the progression of the defects to be monitored.

Many lessons were learned from this first industrial operation. Conclusions should usefully be drawn from these lessons and applied to future developments in inspection equipment. At the same time, laboratory tests have demonstrated the advantages of acoustic emission for monitoring the tendons of tension leg platforms. A relatively major development of this technique is foreseeable in this direction.

Vibrational Examination

This has experienced considerable development over recent years. Based initially on measurements of vibration in the decks of structures under ambient excitation, the lack of results led this method to develop subsequently into two complementary branches: modal analysis, involving forced excitation of the platform on the one hand, with the measurement of flexibility under natural excitation but with extremely advanced signal processing, on the other. These two techniques enable major defects to be detected (weakening of foundations, cut through a member, etc.) on all types of structure. Modal analysis has already been commercially applied in the North Sea, while flexibility measurement is still at the offshore testing stage. A great deal still remains to be done, particularly in this second aspect, but the future does seem open to a technique such as this, to judge by the interest being shown by the American oil companies. It should also be noted that vibrational examination methods will in the near future be the only NDT tools available for inspecting concrete platforms.

Clearly, subsea inspection is an extremely innovative area. And the above list of new techniques is far from exhaustive. New inspection strategies are being investigated, to co-ordinate and optimise the application of these tools.

Their essential advantage is in defining any truly dangerous defect(s) encountered in particular structures, hence providing these structures with an appropriate means of inspection for ensuring that such dangerous defects (which may well be considerably greater than a mini-crack) are duly detected. Once again, there still remains a great deal to be done, but this work is essential if we are to build up a truly efficient subsea inspection industry.

Chapter 10

Non-destructive Testing

Non-destructive testing can be defined as a set of procedures to reveal defects, which may or may not be hidden, in industrial products without significantly modifying the appearance or characteristics of the product.

The inspection of offshore oil installations by non-destructive testing is the 'preventive medicine' for the maintenance of these installations. The inspection of installations is justified for major economic reasons: any lack of knowledge of defects and damage to the installations can lead in the medium or long term to a partial or total halt to production from an installation or possibly from an entire oilfield.

Precisely because of concern about the potential consequences of a lack of inspection and/or maintenance, the governments of many producer countries require a systematic inspection of installations. These inspections are set up according to certification criteria based on the construction of the installations.

When the inspection has been carried out, the state of the installation will be described in a report. Depending on the individual circumstances, the report will be sent to the oil company owning the installation, to the classification body, or to the government body.

In certain cases, no inspection or certification is laid down. Nevertheless, an increasing number of oil companies are carrying out inspections of all their offshore oil installations on their own initiative and at their own cost (see Appendix 24).

The methods of non-destructive testing have been described in the previous chapter on foreseeable developments in inspection techniques.

Chapter 11

Conclusions

In the 'modern' philosophy of plans for marine structures, the inspection, maintenance and repairs should be an integral part of the considerations included in the plans. This is only partially accomplished today, for example in applying particular testing requirements to the construction of a unit incorporating elements which are difficult to inspect during operation. The effectiveness of inspection can be incorporated in the evaluation of structural safety, in the same way that it can be evaluated from an analysis of the structure by an 'advanced' procedure. An example of this latter application has been published*.

The FASTRUDL code is an example of a mathematical model applied to testing procedures. A calculation code such as that recently developed by the CTICM permits the study of the various parameters which govern the behaviour of offshore platforms of the jacket type, by far the most common. This widely used programme gives valuable information on the dynamic response of the structure.

The model shown in the annexes (see Appendix 25) presents several sub-systems connected with the main criteria (load and model for the swell, hydrodynamic forces, ground-structure static interactions, fatigue, cut-outs in the tubes and earthquakes). Such a model should be completed by taking account of the dynamic behaviour of the piles-ground system.

Three types of modelling can be used for studying the pile-ground interaction:

(1) modelling by discrete points with the definition for each of these points of mass, damping and rigidity;
(2) models involving the mechanics of continuous media;
(3) models involving finite elements, which seems to be the best method.

----

* See H. Crohas. (February, 1983) 'Approche probabliste de l'efficacité des inspections par plongeurs sur les plateformes du type jacket'. Bulletin Technique du Bureau Veritas.

Part IV

Technical Appendixes

Appendix 1

Safety in the Nuclear Field

This appendix is designed to demonstrate the full range over which risk analysis methods are used by the CEA. The strengths of the methods used by the CEA are as follows.

- Probabilistic approach to nuclear safety - systems (components and assemblies), accidental sequences, risk assessment.
- Failure Mode and Effect Analysis (FMECA).
- Fault tree assessment - cut-off; minimum cut-off.
- The treatment of states - the Markovian process of transition between successive states.
- Assessment of probabilities of state - Monte Carlo simulation.
- Data bank compilation and use.
- The treatment of uncertainty in data.
- Common mode failures - design, fabrication, erection, human error, environment, etc.
- Accidental sequences which reveal that no accident is the result of a single event or of failure of a single system, taking into account the numerous redundancies.
- Probabilistic studies of risk in a nuclear installation (core melt down) and in the public (deaths) - the Rasmussen report (WASH 1400).
- Comparative analysis of the deterministic approach and the probabilistic approach.
- Analytical criteria.

This appendix provides a detailed insight into functional reliability analysis and the probabilistic approach to risks in the nuclear industry. It includes: (1) an introduction to the role of probabilistic studies; (2) an investigation of the probabilistic approach (systems reliability, accidental sequences and risk assessment); and (3) a summary of the use of probabilistic methods in ensuring the safety of the nuclear industry.

NUCLEAR SAFETY PRINCIPLES - REGULATIONS

Regulations are not based on probabilistic criteria. The criteria are DETERMINISTIC. The probabilistic approach is an aid to safety analysis.

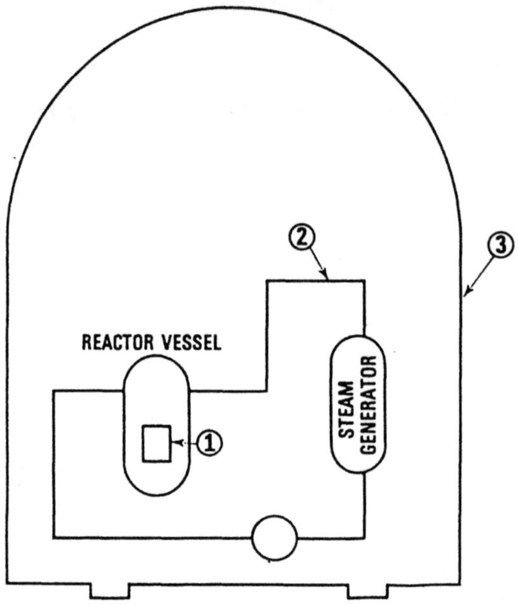

Fig.A.1.1. Deterministic design principle. 'In depth defence concept (three barriers). Reference accident, sizing, accidents; large safety margin; single failure criterion.'

THE USE OF PROBABILISTIC STUDIES

(1) Checking safety levels (no weaknesses).
(2) Choosing between various designs.
(3) Taking certain accidents into account in sizings (external aggression).
(4) Comparing the global risk with non-nuclear risks (WASH 1400).

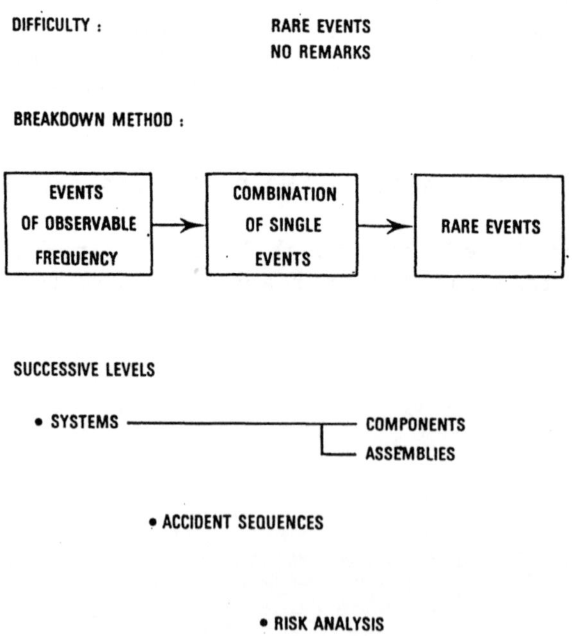

Fig.A.1.2. Probabilistic approach to nuclear safety.

COMPONENT RELIABILITY

Historically: data collected

as a result of safety preoccupations in the

- military
- aeronautical
- nuclear fields

The Concept of Failure

failure — equipment no longer capable of performing its function

- catastrophic failures
- incipient failures

PARAMETERS USED IN COMPONENT RELIABILITY

(A) FAILURE RATE DURING OPERATION

(B) FAILURE RATE ON DEMAND

(C) AVERAGE TIME OF CORRECT OPERATION

(D) REPAIR RATE

Fig.A.1.3.

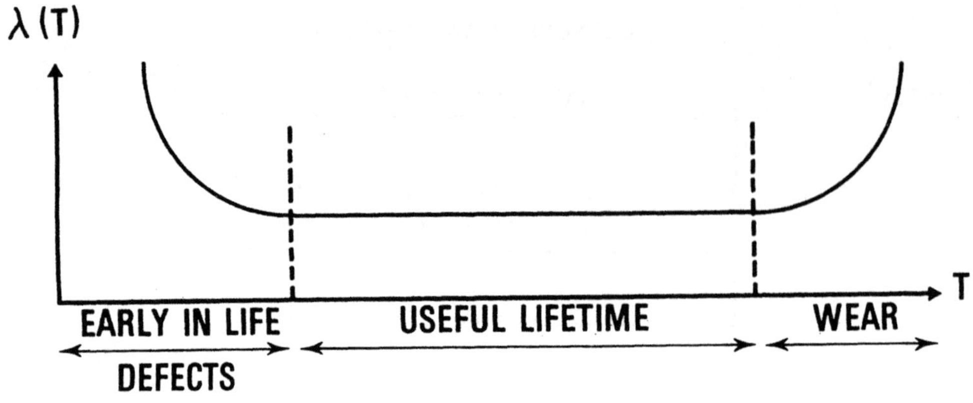

λ is assumed to be non-time dependent

λ - CTE

Exponential law:

- simple calculations
- absence of data

means

- debugging
- preventive maintenance

Estimator: $\hat{\lambda} = \frac{\text{number of faults observed}}{\text{cumulative operating time}}$

Other laws:

- normal log
- Weibull

Fig.A.1.4.

$$\gamma = \frac{\text{number of refusals}}{\text{cumulative number of operations}}$$

- diesel units
- circuit breakers

Mean Time Between Failures

λ = CTE

→ M.T.B.F. = $\frac{1}{\lambda}$

Mean Time to Repair

μ = CTE

↓ M.T.T.R. =

$$\frac{1}{\mu} = \frac{\Sigma \text{ time to repair}}{\text{number of repairs}}$$

SYSTEMS RELIABILITY

Qualitative Assessment

- FMEA (failure modes and effect analysis)
- Fault trees.

Quantitative Assessment

(1) Methods:

- fault trees
- processing of states (Markov)
- Monte Carlo simulation

(2) Reliability data:

- data banks
- common mode failures
- human error

ANALYSIS OF COMPONENT FAILURES AND THEIR EFFECTS UPON THE SYSTEM

FMEA (failure mode effect analysis)

Description of the Method

- Inductive method.
- For each component the following must be defined:

(a) the system's function;
(b) its modes of failure;
(c) the possible causes of failure;
(d) the effects on the system;
(e) the means of detecting failure.

Advantages of the Method

- This is a rigorously systematic and efficient method of analysis of a system's failures.
- Identification of simple failures.

Disadvantages of the Method

- Complex systems present redundancies: the method is long and fastidious.
- Human errors and the effects of a difficult environment must be taken into account.

FAULT TREES

Definition

A logic diagram showing the relationship between the undesired event (system failure) and the elementary events liable to cause it (component failure).

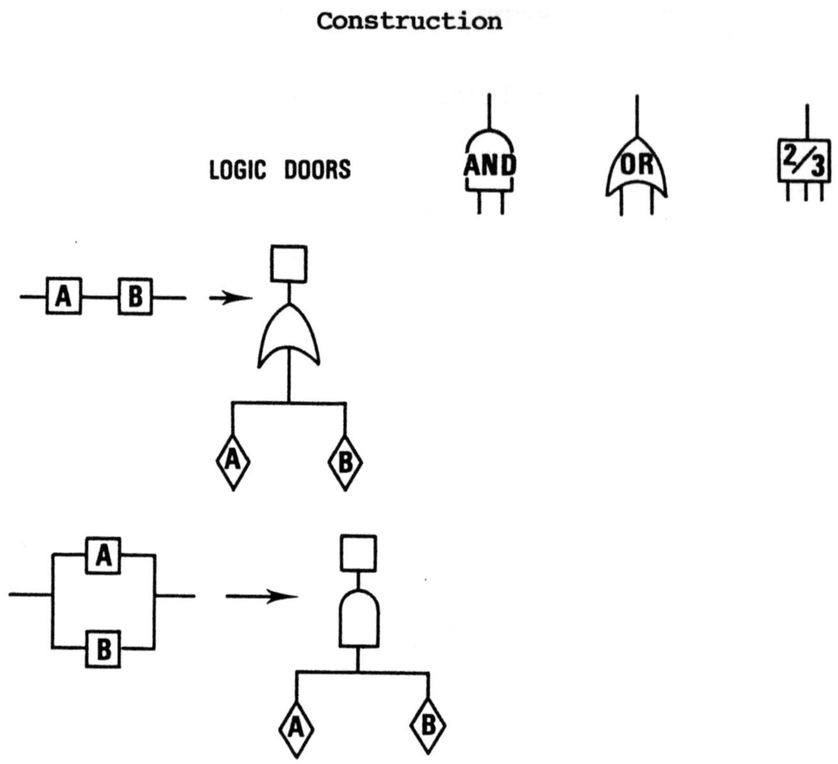

Fig.A.1.5. Construction of a fault tree.

Table A.1.1. Failure mode and effect analysis (FMEA)

Component (Tag. No., name, type, location)	Function states	Failure modes	Possible causes of failure (internal/ external causes)	Effects on the system	Means of detection	Comments
Tag No: 210 VB Name: emergency boron valve Type: motorised on-off globe valve	Function: to bring in emergency boron feed Status: position controlled from control room on basis of indications on flow meter 60 mD when emergency boron in use. - otherwise closed.	- jammed open - jammed partially open	- boron crystal-lisation - seizing - motor failure	- impossible to use emergency boron line (60 mD) - impossible to achieve adequate flow of boron	- boron flow rate in emergency boron line	- failure of this item was not considered in the quantitative analysis as its use is exceptional only

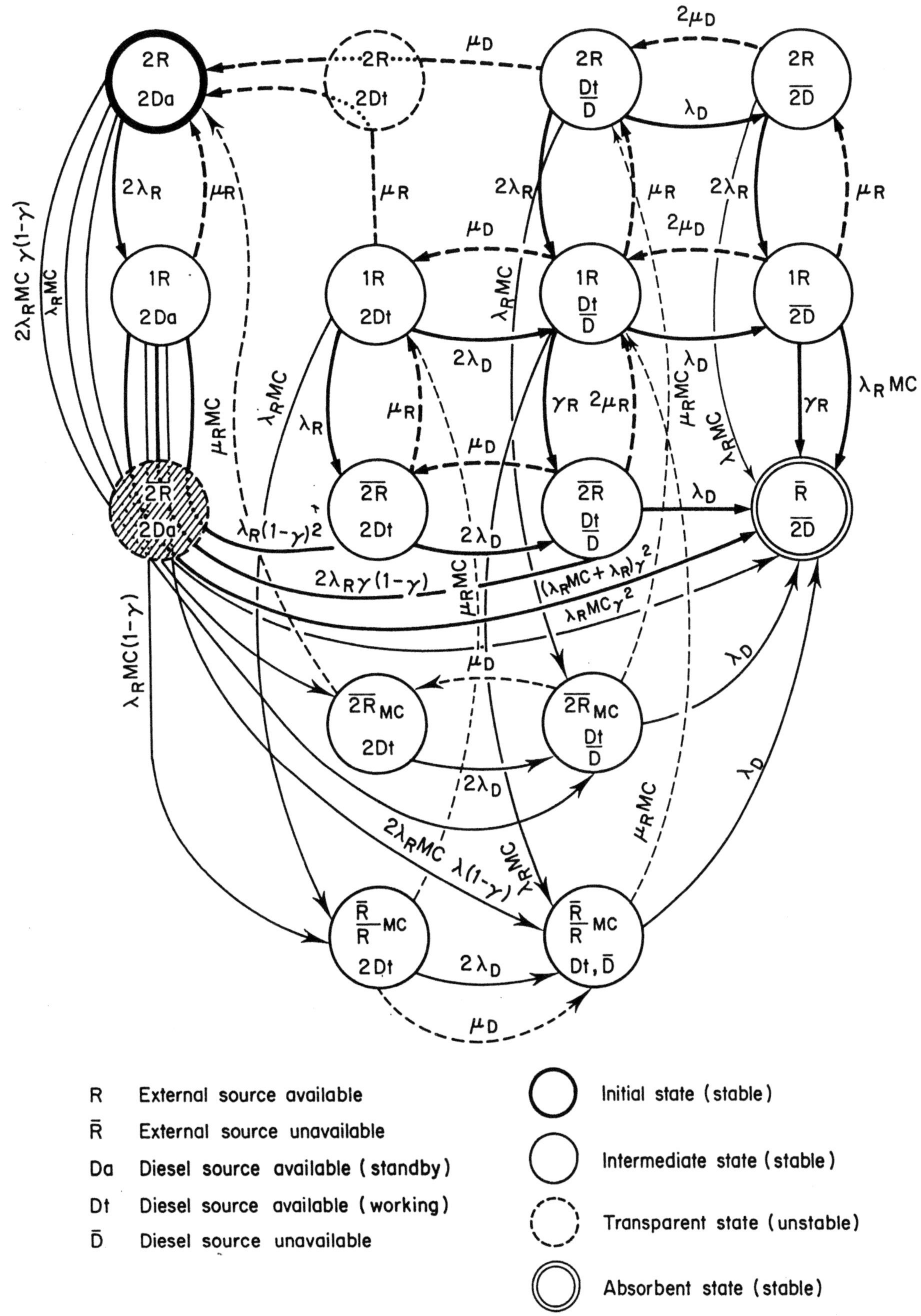

Fig.A.1.6. Definition of states and transition between states.

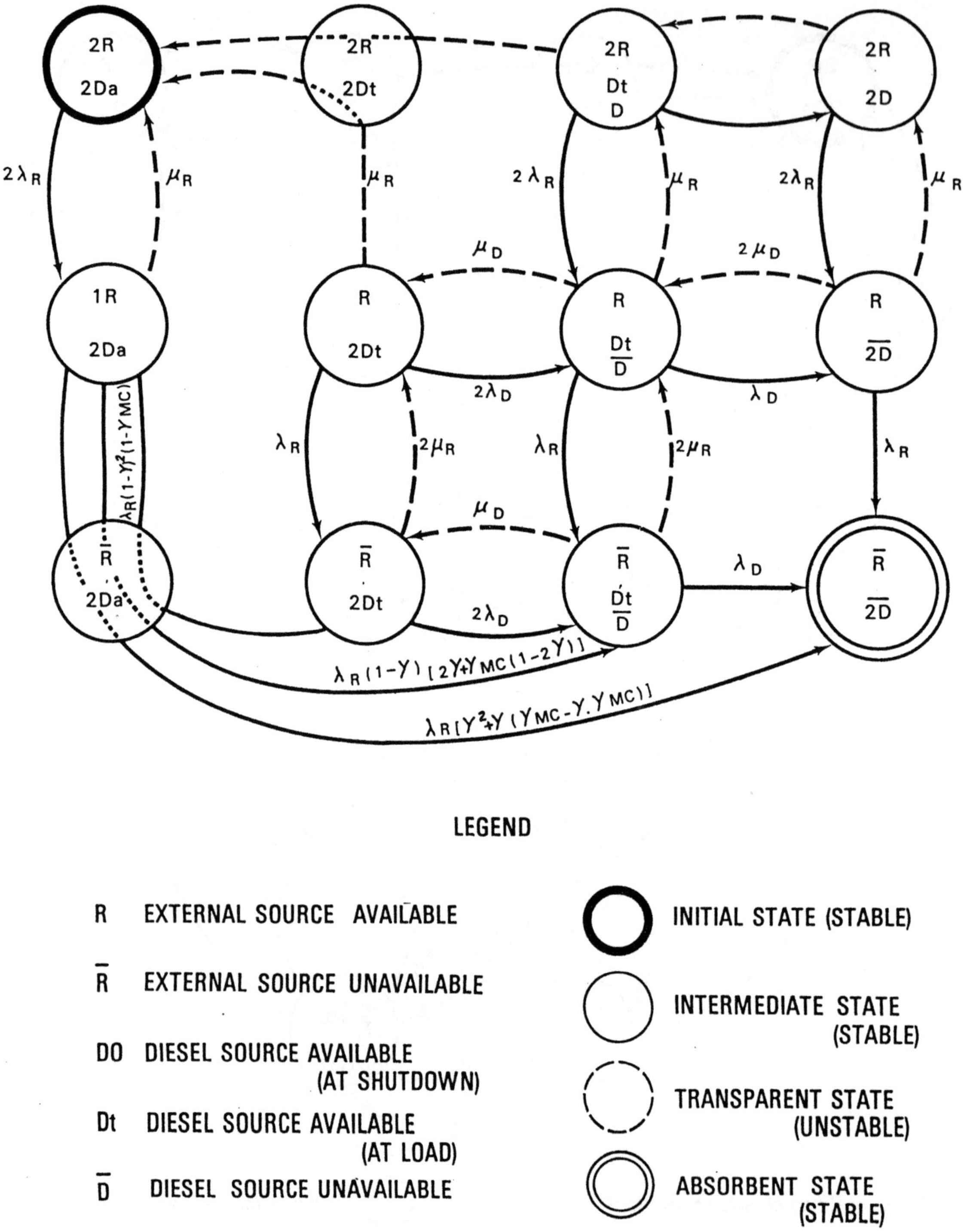

Fig.A.1.7. Definition of states and transition between states.

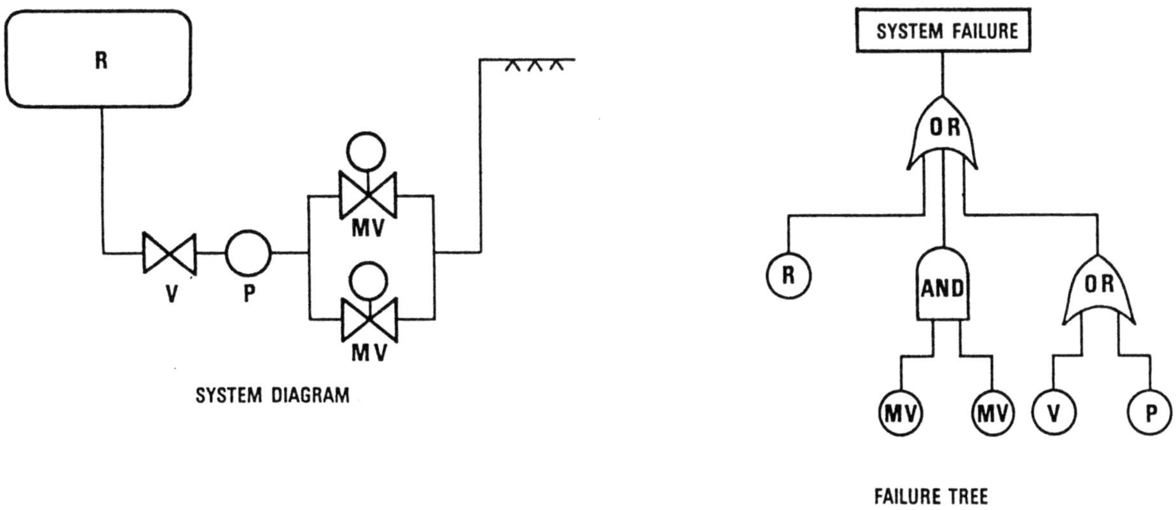

Fig.A.1.8. Fault trees.

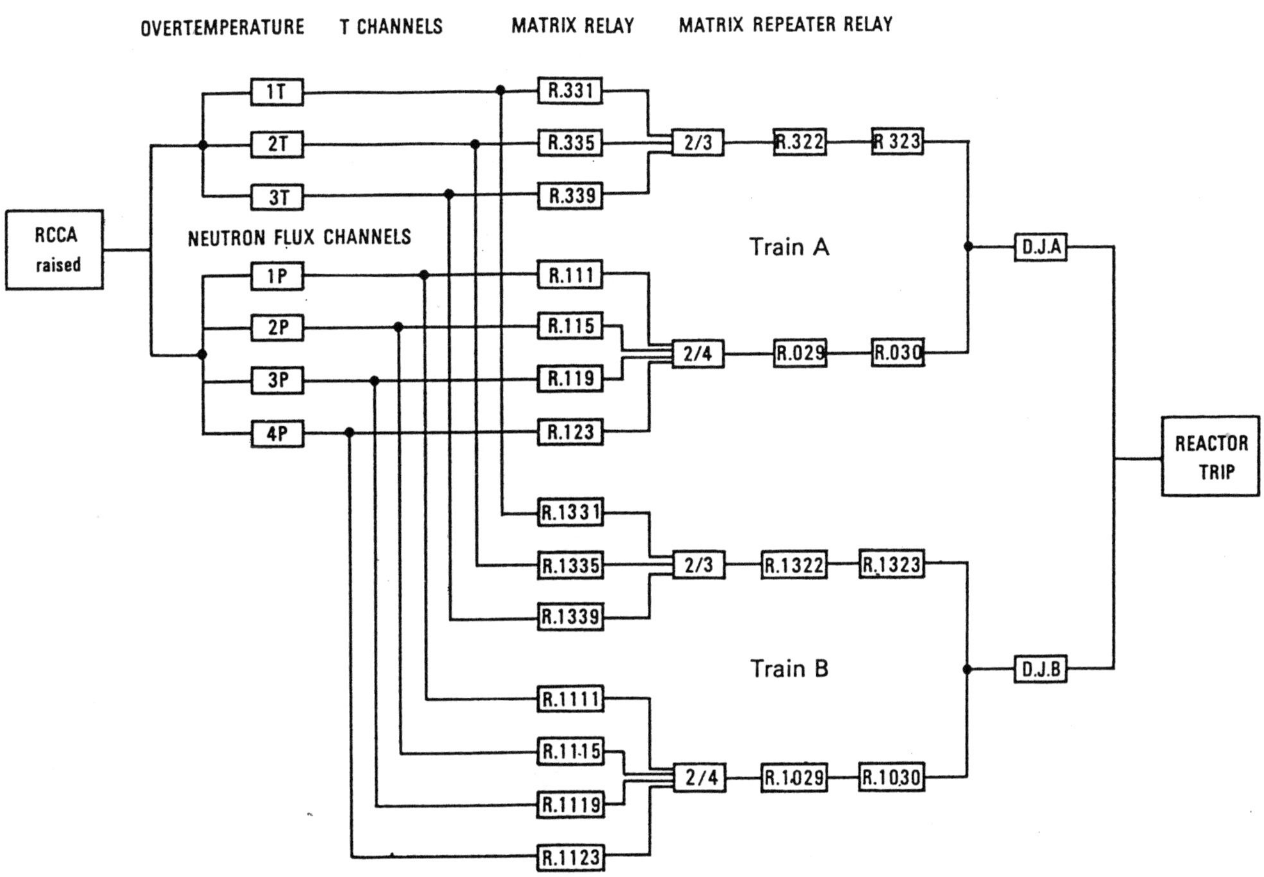

Fig.A.1.9. Sequence diagram.

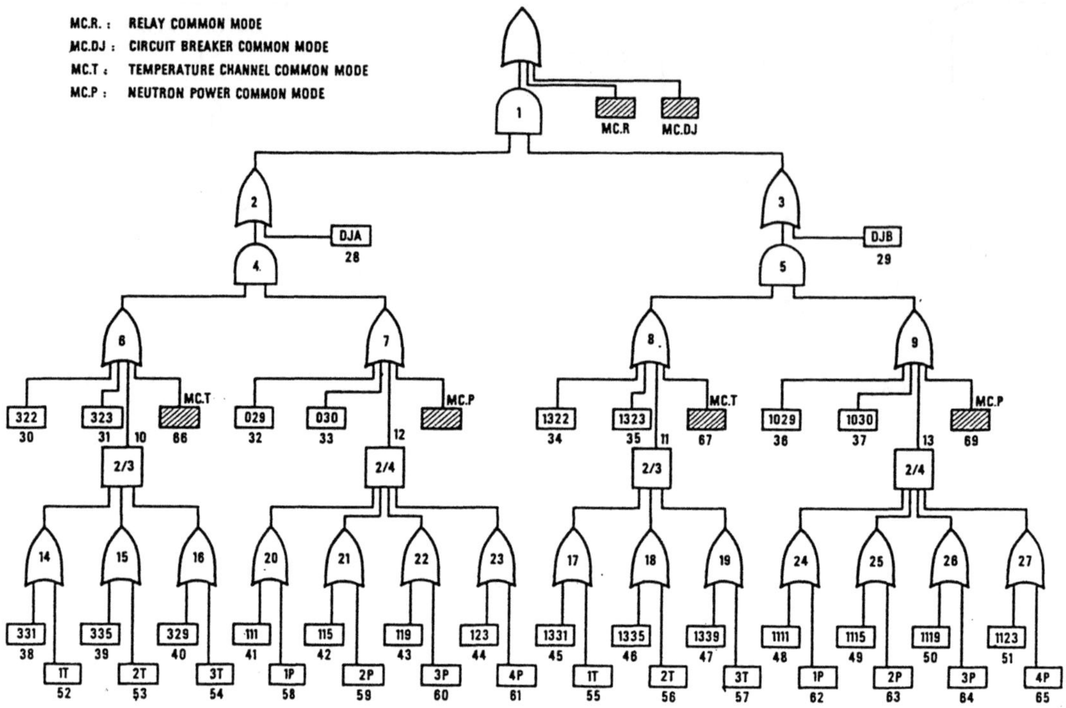

Fig.A.1.10. Fault tree.

FAULT TREES CUT-OFF PATHS METHOD DEFINITIONS

Cut-off:	set of components in which simultaneous failure involves failure of the entire system.
Order:	the order of a cut-off is the number of components in the cut-off.
Minimum cut-off:	occurs if not containing a cut-off of a lower order.

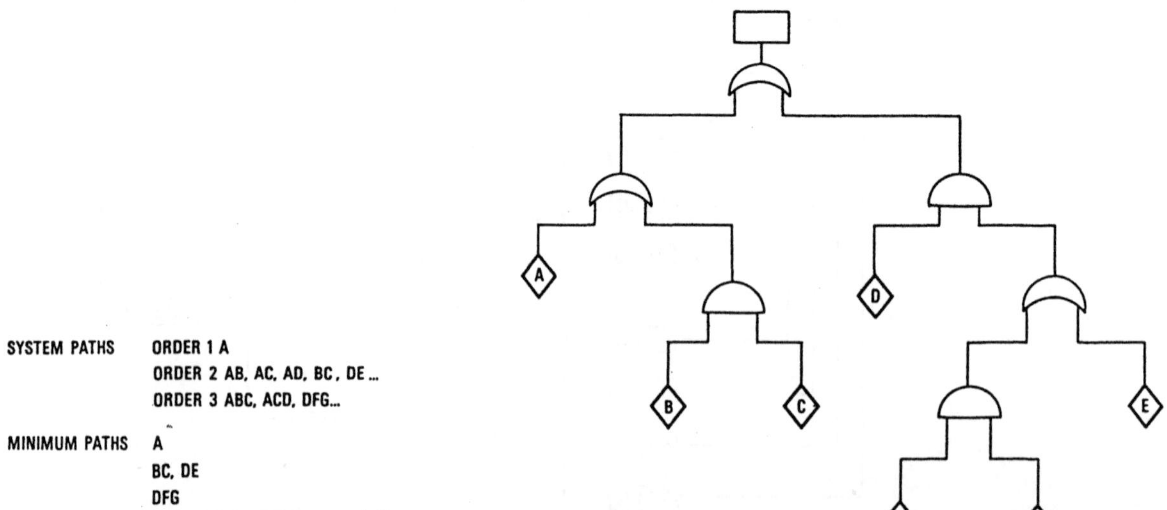

Fig.A.1.11. Fault trees. Cut-off paths method - examples.

FAULT TREES CUT-OFF PATHS METHOD USE OF THE METHOD

Qualitative Interest/Quantification

$$PT = P\left(\bigcup_I C_I\right) = \sum_I P(C_I) - \sum_{I<J} P(C_I C_J) + \ldots$$

$$= \sum_I P(C_I)$$

COMPUTATIONAL CODES

PREP and KITT
SETS
FTAP
FIABC

FAULT TREES CUT-OFF PATHS METHOD. LIMITATIONS OF THE METHOD

The method is applied when:

(1) components have only two states;
(2) the elementary laws of probability apply.

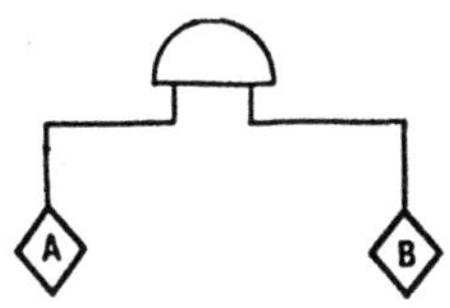

P = PA . PB

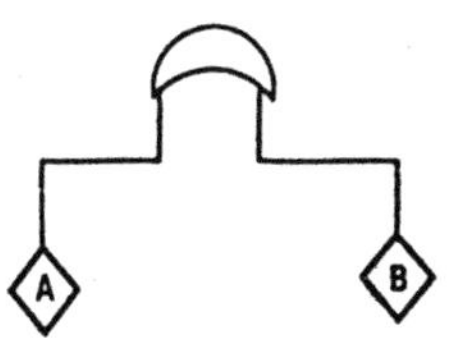

P = 1 - (1-PA)(1-PB)

e.g., reliability of non-repairable systems, refusal to start, and availability of repairable systems.

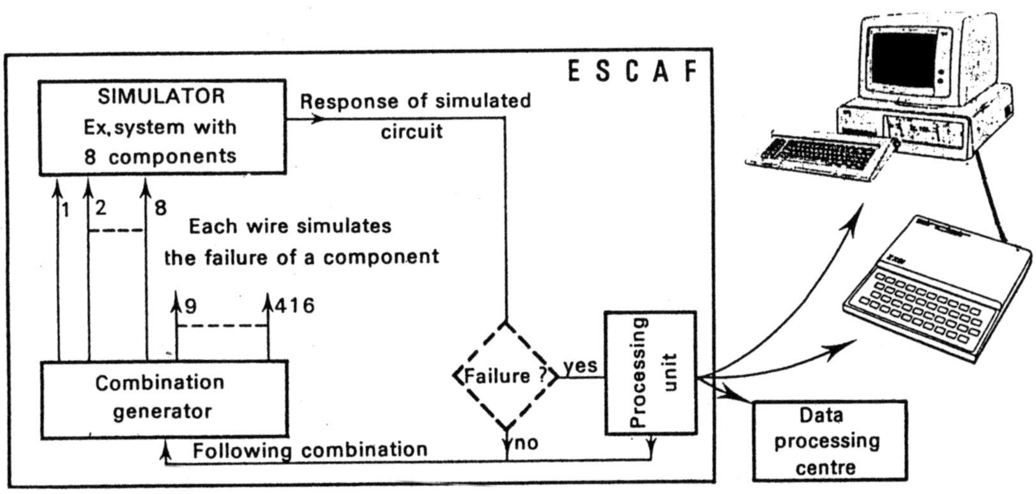

Fig.A.1.12. ESCAF

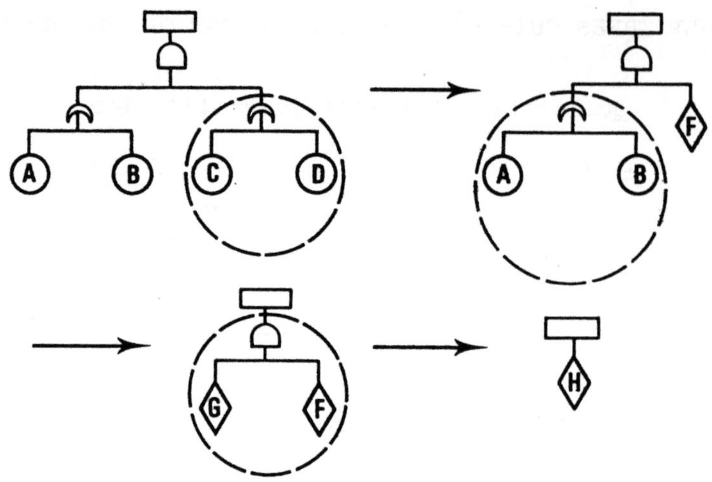

ADVANTAGE : NO APPROXIMATION

DIFFICULTIES : CASE OF A LARGE NUMBER OF DEPENDENCES

DESIGN CODES : PATREC, PLMOD

Fig.A.1.13. Fault tree, direct computation. The principle is to progressively replace the sub-trees by their pre-tabulated values.

TREATMENT BY STATES
THE MARKOV METHOD

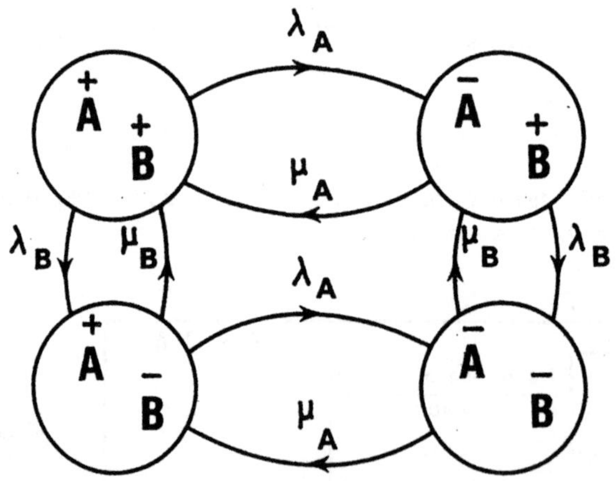

Fig.A.1.14. Markov method - the principle.

$P_I(T)$ is the probability of being in state I at instant T and conforms to a differential equation of the form:

$$\frac{DP}{DT} = A.P. \qquad P = \text{vector, states PI} \qquad A = \text{transition matrix}$$

$P_I(T)$ may be calculated numerically when A is not time-dependent (exponential laws).

Applications

- Reliability of repairable systems.
- Components with more than two states (stand-by).

Codes

MARK.GE - FIABI

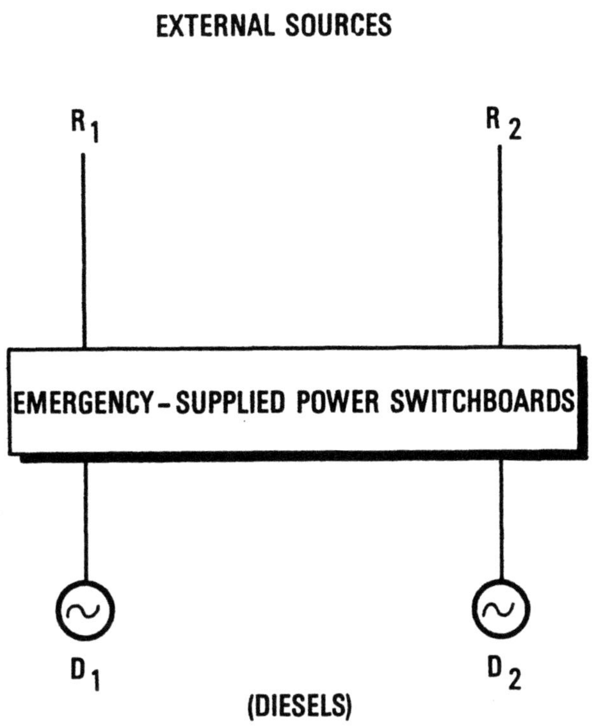

Fig.A.1.15. Treatment of states. Example: electricity supply flow chart.

MONTE CARLO SIMULATION

Principle

(1) Computer simulation of system's 'case history' by random sampling (e.g. A breaks down at time T_A; B breaks down at time T_B; A is repaired at time T_A, etc.).

(2) Results obtained statistically following an adequate number of case histories.

(3) Need to speed up calculation: biassing.

Applications

Applications make it possible to take into account:

- the order of failures;
- the duration of failures;
- and as a function of time, etc.

Computational codes: MONARC, BIAF.

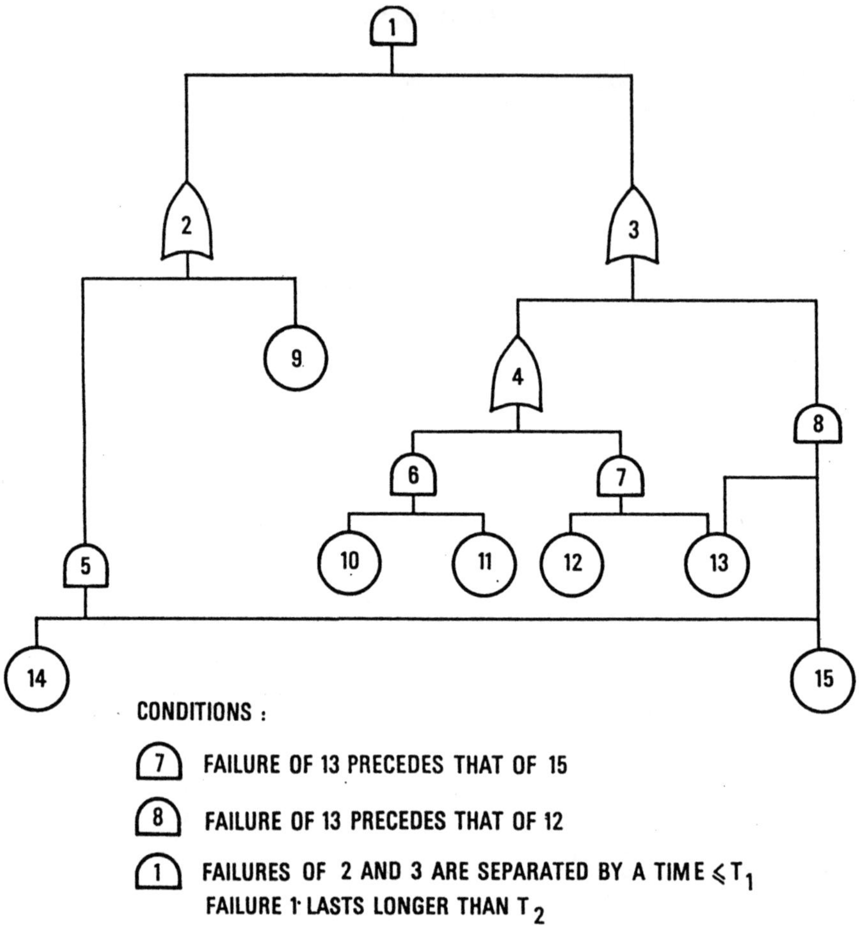

Fig.A.1.16. An example of Monte Carlo simulation.

DATA BANKS

Non-nuclear

- electronic - CNET data compilation (electronic components, relays, etc.).
- electromechanical and mechanical - NPRD1, 1978.
- syrel data bank (UK) (source: various industries).

Nuclear

- WASH 1400 appendix (compilation of existing data)
- IEEE standard 500 - electrical and electronic equipment in nuclear power stations.
- NPRDS - American nuclear power station equipment, but heterogeneous due to large number of operating companies.
- Phenix file (AEC) sodium components.
- EDF data banks - SRDF, 1978:
  - over 6 PWR sections (Fessenheim and Bugey)
  - 800 equipment items (electrical and mechanical).

TREATMENT OF UNCERTAINTY IN DATA

(1) Precision on estimate of a failure rate (λC^I) ====> rule of χ^2 ====> confidence gap. This is dependent on: (1) number of failures, and (2) cumulative operating time.

(2) Dispersion of data due to heterogeneity of sources

WASH 1400 method - lognormal law:

- mean value for λ;
- markers in confidence gap (5-95%);
- error factor.

(3) Treatment of uncertainties on data, in computational codes

- analytical method (propagation of variance)
- Monte Carlo method

====> confidence gap on the final result.

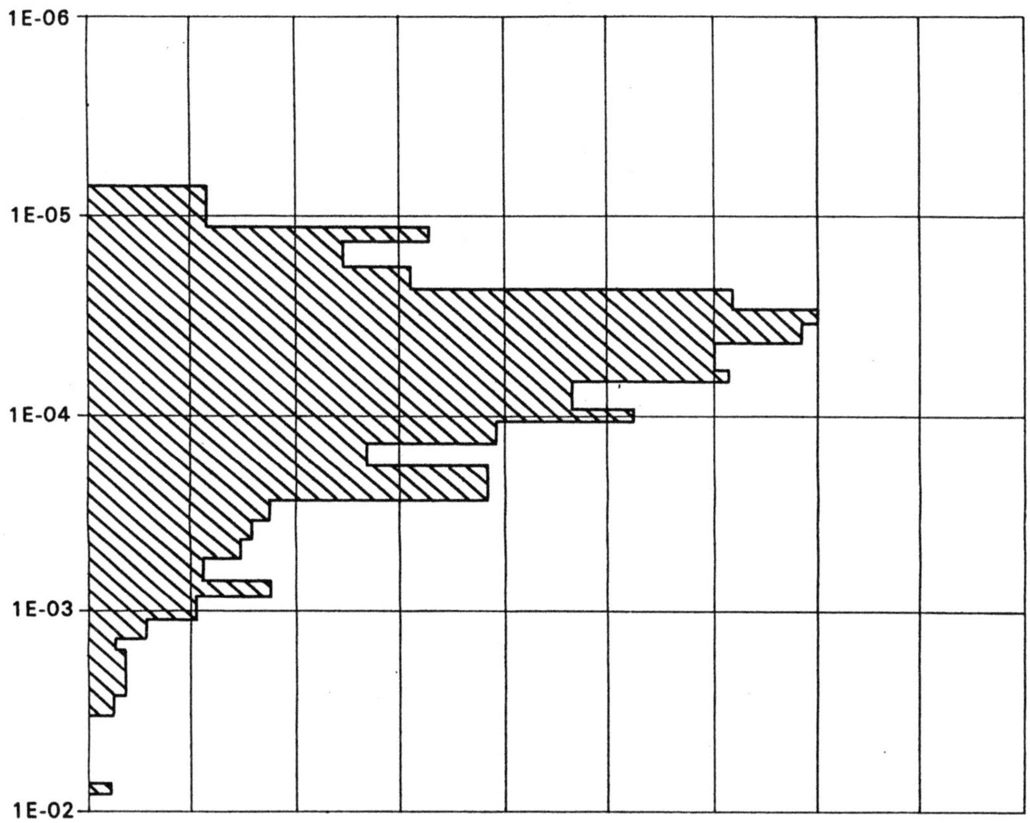

Fig.A.1.17. Distribution histogramme.

COMMON MODE FAILURES

Definition

Multiple failures due to a common cause.

Classification

- Design errors:
  - (a) functional deficiency
  - (b) unsuitability for the intended purposes
- Manufacturing errors
- Procedural errors:
  - (a) operating errors
  - (b) testing or maintenance errors
- Environmental effects:
  - (a) normal environment
  - (b) extreme environment
  - (c) energy events

Table A.1.2. Example of classification of common causes of failure

COMMON CAUSES OF FAILURE							
Technical				**Operational**			
Design		**Manufacture**		**Procedures**		**Environment**	
Functional deficiencies	**Performance errors**	**Factory**	**Installation and control**	**Operations**	**Maintenance and testing**	**Normal and extreme**	**Energy events**
unforseeable risk	inter-dependent channels	inadequate quality control	inadequate quality control	human error	faulty repair	temperature	fire
inadequate instrument-ation	operational deficiencies	inadequate specifi-cations	inadequate specifi-cations	inadequate procedures	faulty testing	pressure	flood
inadequate control	inadequate components	inadequate inspection	inadequate inspection	inadequate inspection	faulty calibration	humidity	storm
	design errors	inadequate testing	inadequate testing and control	communication error	faulty procedures	vibration	earthquake
	limitation in design				faulty control	acceleration	explosion
	common protection items					voltage	missiles
						corrosion	radiation
						contamination	chemical sources
						interference	short-circuit
						radiation	
						static charge	

COMMON MODE FAILURES - EXAMPLES

Design Errors

Zion 2, 1974 (SGES) - air pockets at high points.
Zion, 1974 ('RIS') - errors in wiring diagrams and unsuitable commissioning tests.

Erection Errors

Point Beach, 1974 (SGES) - filters installed for tests and not subsequently removed - plugging.

Human Errors in Maintenance

Turkey Point, 1974 (SGES) - inadequate tightening of seals on three pumps - impossible to start up.

Environmental Effects

Kewaunee, 1975 - filters blocked by grains of resin.
Palissade, 1976 - 'PTR' vent blocked by ice.

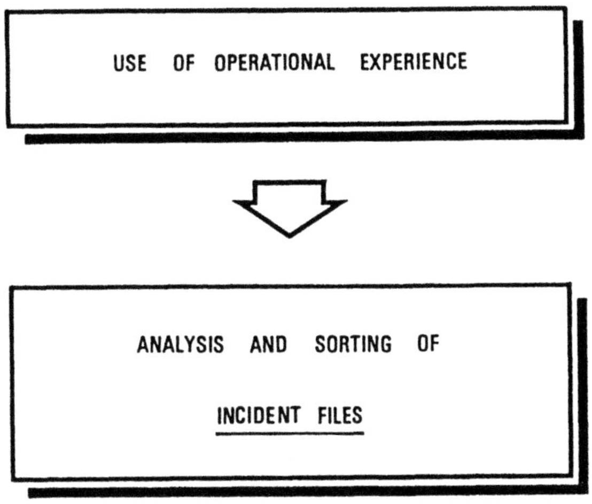

Fig.A.1.18. Common mode failures. Tentative quantification.

COMMON MODE FAILURE - INCIDENT FILES

SAFETY FILES

American Files

(1) From LER (licensee event reports);
NSIC (nuclear safety information centre file).
(2) NPE (nuclear power experience) to more detailed descriptions.

French Files

(1) NSIC file.
(2) DSN file - three categories of incident, according to gravity.
(3) SAPIDE file - more detailed records of the most important incidents.

Computerised files - automatic retrieval system.

AVAILABILITY FILES (EDF)

Operation/Maintenance Analysis Report (OMAR)

This comprises two parts:

(a) An American file, produced by Westinghouse from the NRC Gray Book, the AECH (atomic energy clearing house) and the NPE;

(b) French file according to EDF monthly operating reports.

Rapid Data Retrieval System (RADAR - F)

Developed by Framatome on the model of Westinghouse's system: rapid return of information (weekly) which gives maintenance recommendations.

THE FUTURE IN FRANCE

The new EDF file is computerised into two parts:

(1) events file - all incidents relating to safety and availability;
(2) significant incidents file - extracted from the above file and supplemented by safety analysis.

BREAKDOWN OF COMMON MODE FAILURES

Design errors:	50%
Fabrication and erection errors:	10%
Human errors:	30%
Environmental effects:	10%

Table A.1.3. Example of frequencies observed (American files).

System	Number of years x reactors	Number of faults	Number of common mode faults
SGES	69	75	14
'RIS'	54	220	21

Table A.1.4. SGES.* Typical influence of common mode failures on a system.

	2 pumps	3 pumps
Probability of failure on demand (from files)	4×10^{-3}	2×10^{-3}
Probability of independent and random failures (forecast)	2×10^{-5}	7×10^{-8}
WASH 1400 estimate		3×10^{-5}

However, real unavailability time is generally unknown in the files and gives a pessimistic estimate.

---

* SGES = steam generator emergency supply.

→ METHOD OF FACTOR β

β = (FOR ONE COMPONENT) :

$$= \text{RATIO} \frac{\text{NUMBER OF COMMON MODE FAILURES}}{\text{TOTAL NUMBER OF FAILURES}}$$

Fig.A.1.19. Method of quantifying common mode failures in probabilistic assessments.

HUMAN ERROR

Classification of human error involves the following:

- simple execution tasks (elementary tasks in operation or maintenance);
- complex tasks (may be broken down into elementary operations);
- elementary vigilance tasks (operation);
- complex inspection tasks (maintenance);
- analytical and diagnostic tasks (operation).

Quantification (complex tasks)

Breakdown (THERP) Method

(1) Break down into elementary tasks.
(2) Identify the possibilities of human error.
(3) Quantify the various elements.
(4) Combine the various probabilities and estimate the global value.

Difficulties in Quantifying

- Data banks for elementary operations (but often in laboratory tests).
- Difficulty in quantifying interactions between elementary tasks (breakdown of complex tasks).
- Impossible to quantify analytical or diagnostic errors, or random errors.
- Human error frequently associated with design faults or inadequacies.

To Obtain Better Data

- Use experience data (incident files).
- Use training simulators.

Typical Rates Applied (WASH 1400)

- Incorrect selection of button of a different form (or position) to the button required	10^{-3}
- Error in interpretation (label incorrectly read, hence incorrect button selected)	3×10^{-3}
- Error by omission (forgetting to return a manual valve to its initial position after maintenance - no indication in control room)	10^{-2}
- Error in cases of high stress level	0.2-0.3

ACCIDENTAL SEQUENCES

Taking into account the numerous redundancies, no accident is the result of a single event or the failure of a single system. There has to be a chain of undesirable events, or an accidental sequence.

Probabilistic Assessment

To assess the probability of an accident one has to:

(1) determine the events making up the accidental sequence;
(2) evaluate the corresponding probabilities.

The method involves the build-up of event trees.

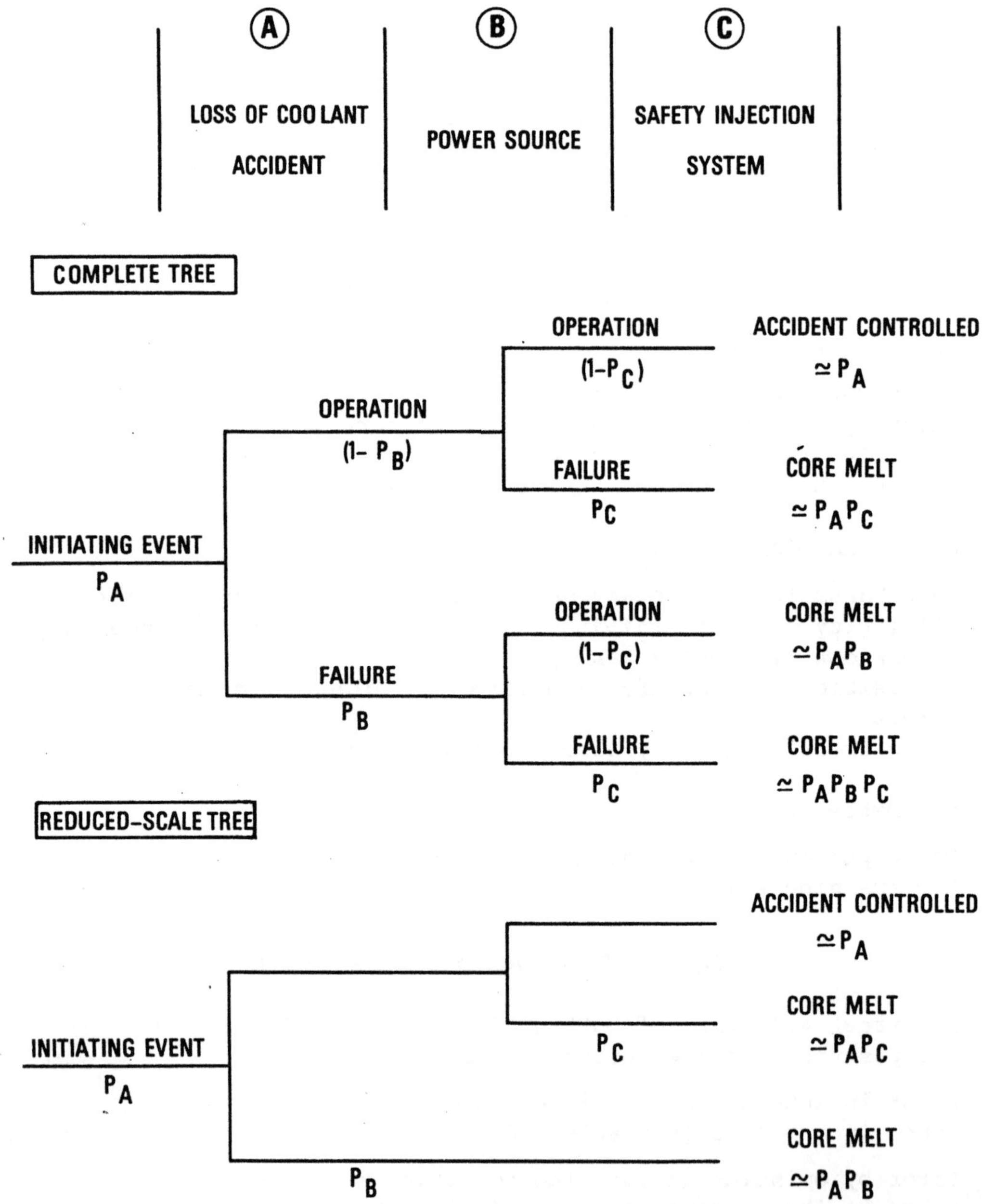

Fig.A.1.20. Accidental sequences. Typical events tree.

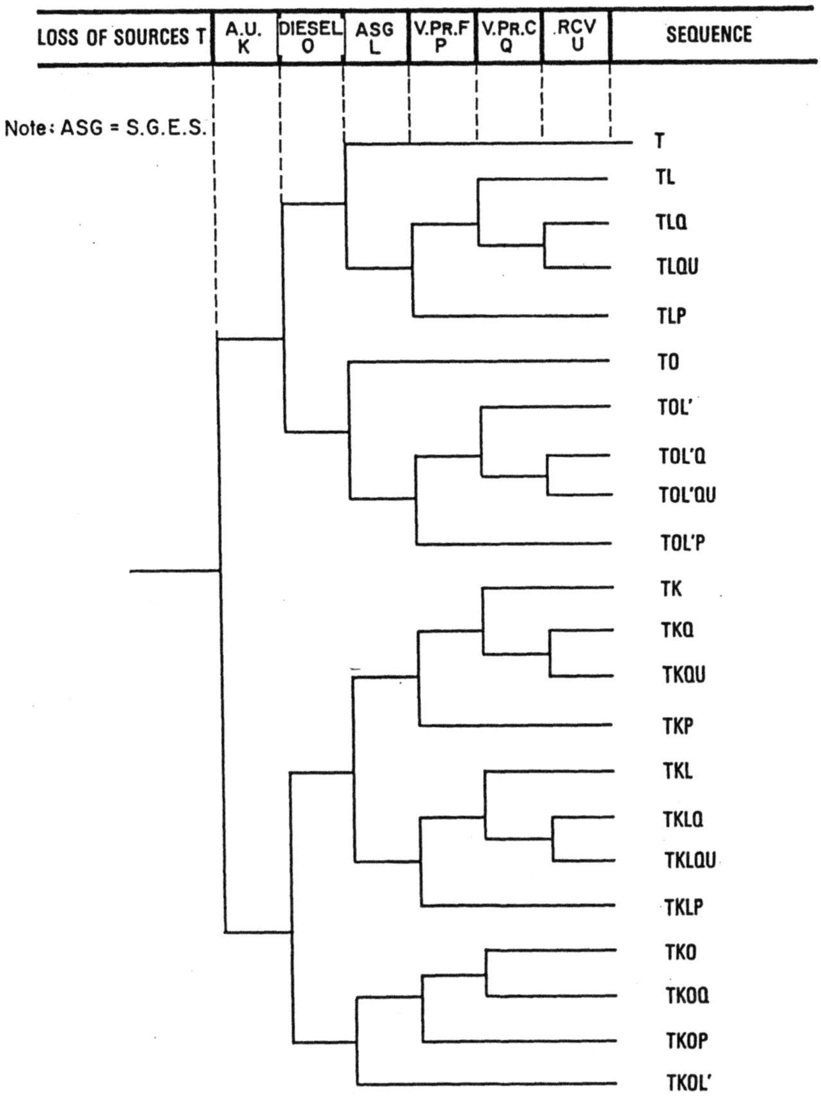

Fig.A.1.21. Accidental sequences. Loss of electricity supply.

PROBABILISTIC RISK ASSESSMENT (PRA)

These are the various possible levels of risk definition:

(1) at installation - risk of core melt down;
(2) among public - risk of more than N deaths.

Methodology

(1) Identify possible initiating events and the probability of their occurring.

(2) For each, draw an events tree as a function of the safeguards provided.

(3) Study the consequences on each branch.

(4) Add the probabilities from the branches to the unacceptable consequences.

Risk Study - An Example

WASH 1400 Report

(1) Research all the imaginable scenarios.
(2) Elimination and regrouping (dominant sequences).
(3) Study of the consequences (classification by category).
(4) Quantification:
  (a) probabilities of accidental sequences;
  (b) consequences (number of deaths).

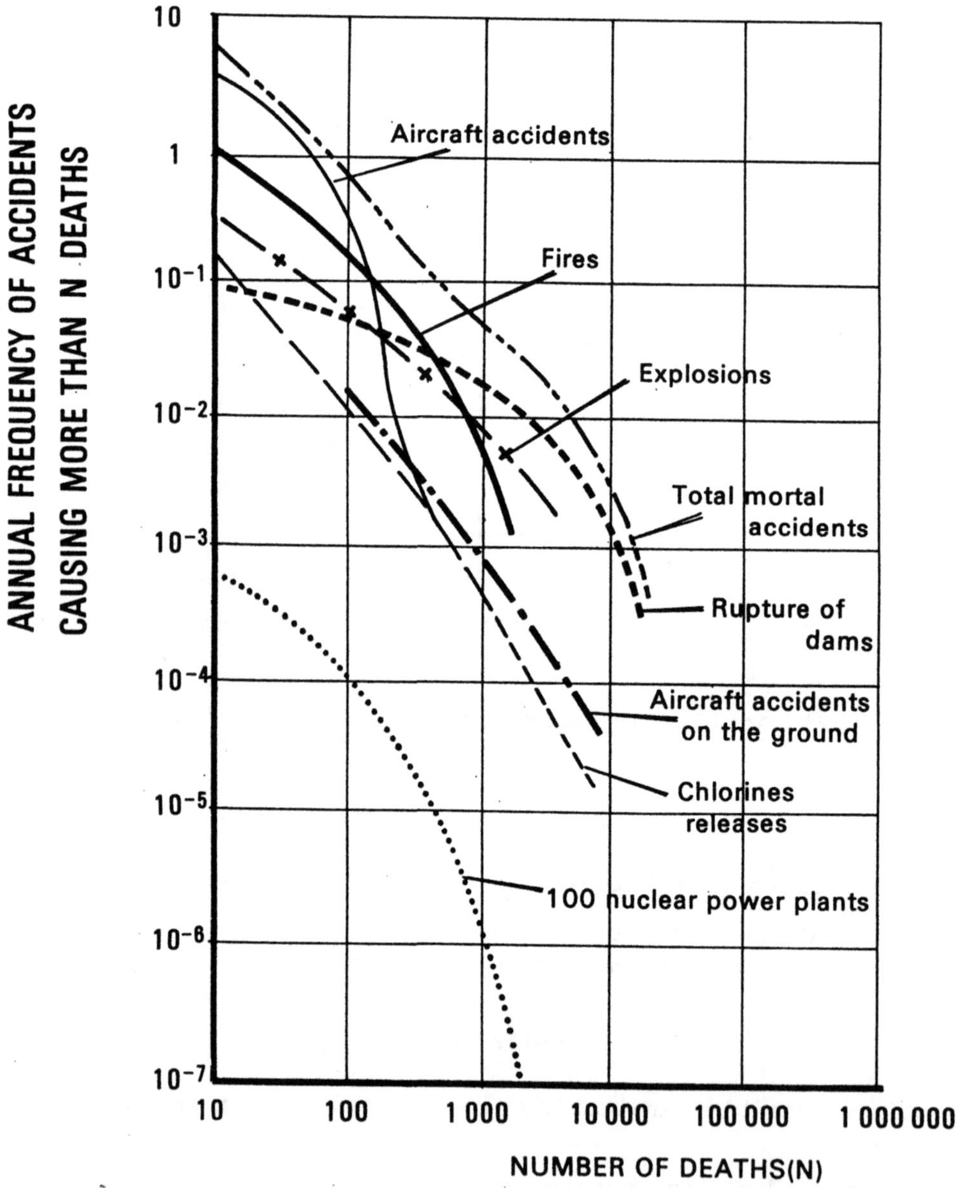

Fig.A.1.22. Risk assessment - an example. A comparison between the risks of nuclear accidents and other risks run by society as a result of human activities.

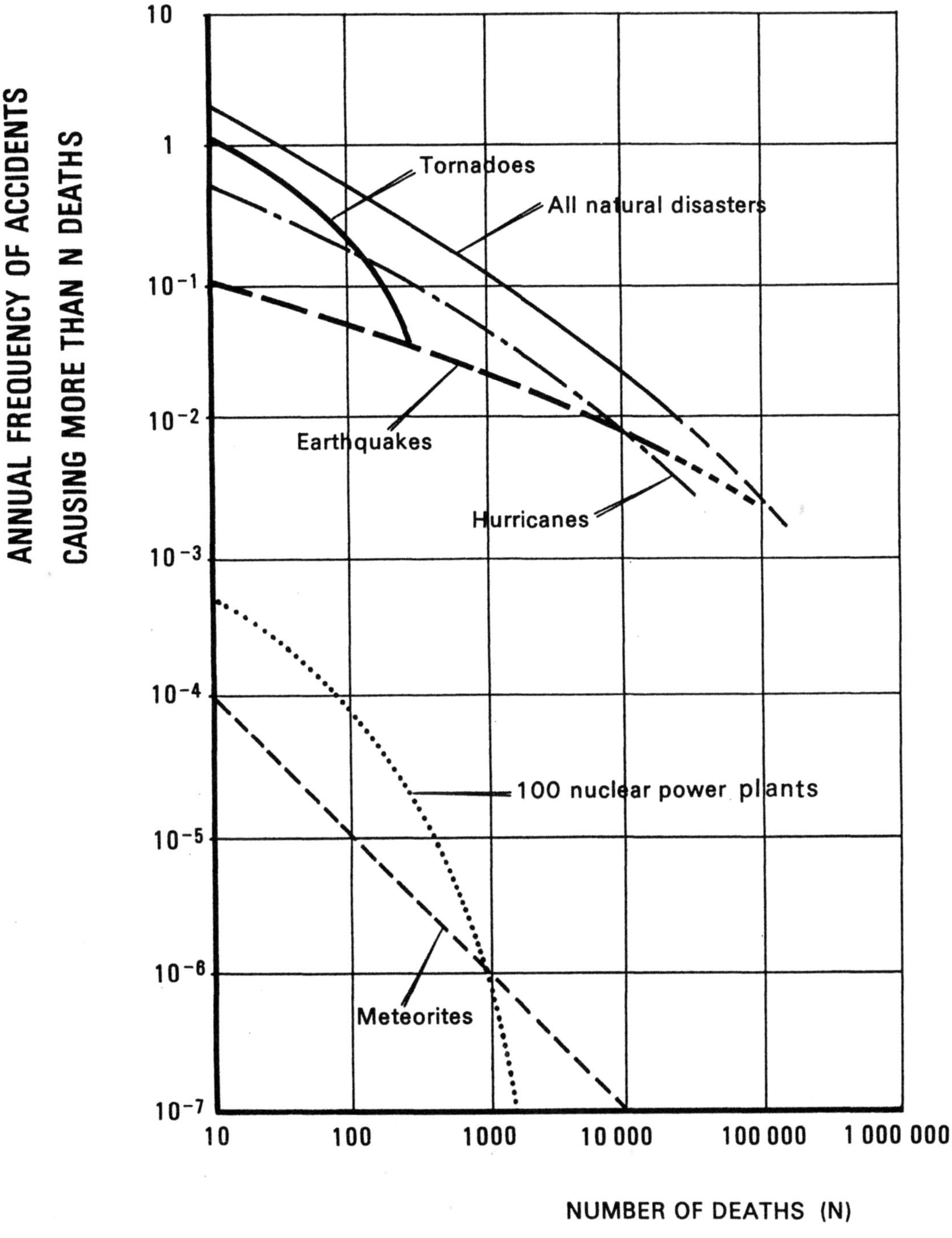

Fig.A.1.23. Risk assessment - an example. A comparison between the risks of nuclear accidents and other natural risks run by society

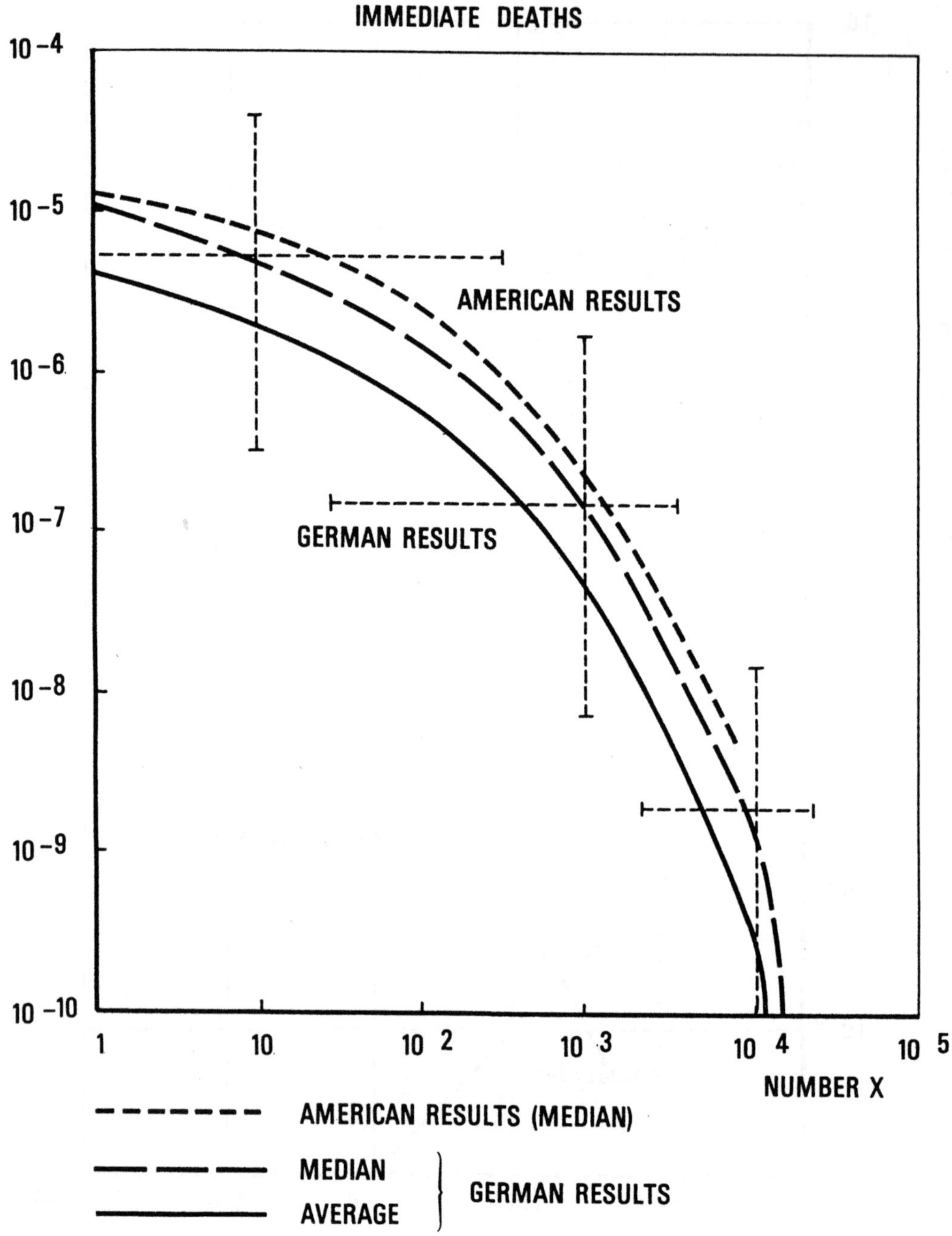

Fig.A.1.24. Comparison between German risk assessment and WASH 1400.

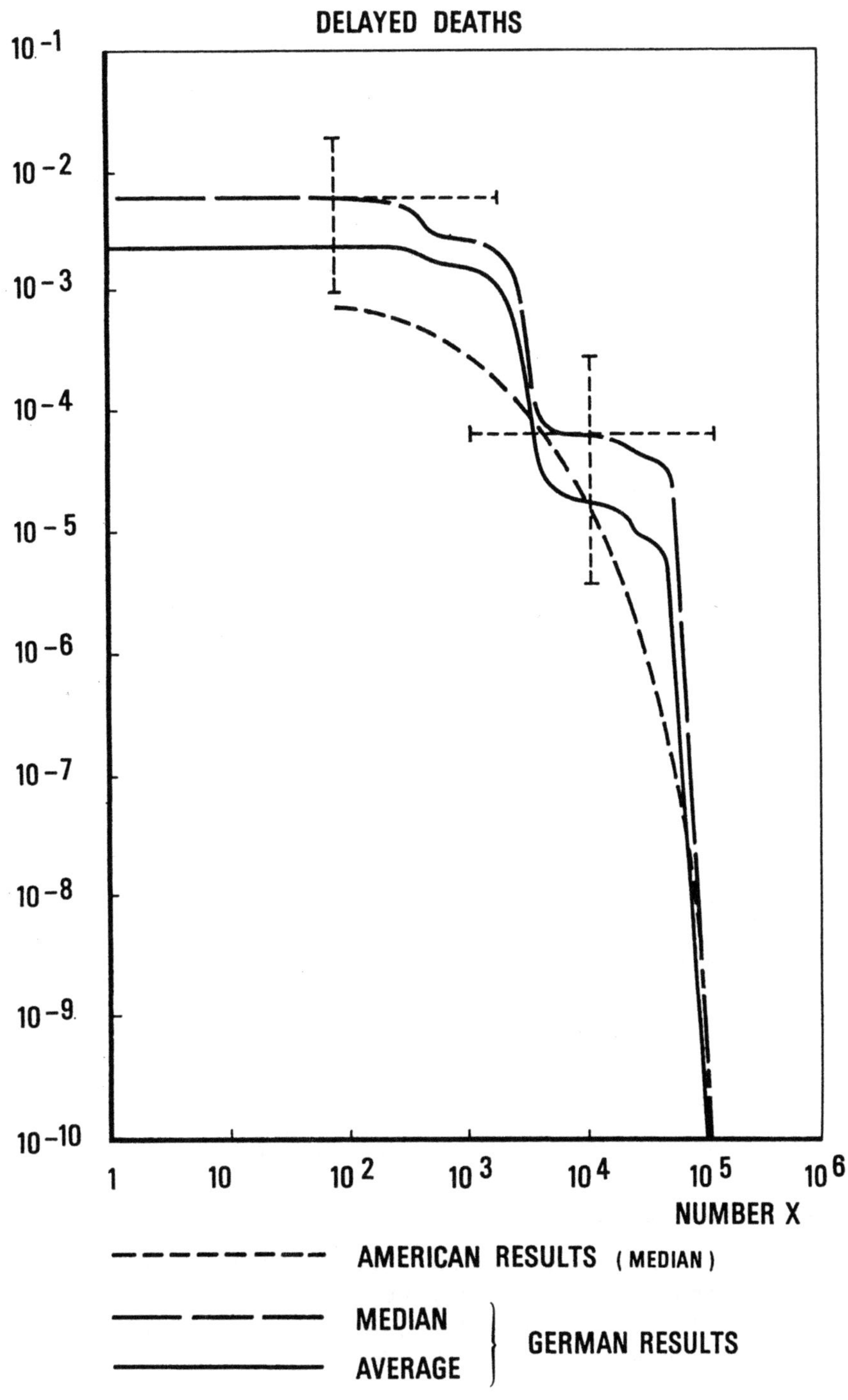

Fig.A.1.25. Comparison between German risk assessment and WASH 1400.

THE USE OF PROBABILISTIC METHODS IN NUCLEAR SAFETY

The current approach is deterministic.

criteria: (a) reference accident
(b) engineering/sizing accidents
(c) single failure

The totally probabilistic approach is exhaustive and realistic but there is the disadvantage of the need to define criteria.

Table A.1.5. Comparison between deterministic and probabilistic approaches.

Deterministic approach	Probabilistic approach
(1) A priori definition of engineering/sizing accidents	(1) Evaluation of the full spectrum of possible accidents
(2) Evaluation of hypothetical and physically non-coherent engineering/sizing accidents	(2) Evaluation of possible and physically coherent accidents
(3) No analysis of accidents with core melt down	(3) Core degradation taken into account
(4) No evaluation: - of the effect of unavailbility associated with testing and maintenance - of human error	(4) These factors more realistically taken into account

APPLICATION OF PROBABILISTIC CRITERIA IN FRENCH REGULATIONS

For PWRs, the SCSIN has stipulated:

(1) global limit probability to be taken into account in engineering/sizing: 10^{-6} reactor-year;
(2) families of events are to be taken into account if the probability of unacceptable consequences is $> 10^{-7}$/year.

However, the limits are regarded as objectives of an indicative nature.

Practical Applications

(a) Risk targetting:
- air crashes
- turbine projectiles: --> 10^{-7}/year-section.

(b) Studies in progress on:
- earthquakes
- floods.

AMERICAN CRITERIA PROPOSED BY THE NRC

The risk of immediate death as a consequence of an accident among the population located within a radius of 1.6 km of a power station must not exceed 1/1000th of the sum of the risks of accidental death of the average American.

The risk of cancer caused by accident in the population located within a radius of 80 km of a power station must not exceed 1/1000th of the total risk of death by cancer in the American population.

To come below the above limits, industry must undertake any modification costing less than $US1000 per man X REM saved.

THE USE OF PROBABILISTIC METHODS WITHOUT A GLOBAL CRITERION - RELATIVE ASSESSMENTS

(1) Full risk evaluation related to a reference power station.

(2) Relative evaluations in relation to this reference in order to eliminate (1) criteria definition, and (2) some of the uncertainties.

APPLICATIONS OF RELATIVE ASSESSMENTS

(1) Existing power stations:
  (a) importance of modifications;
  (b) gravity of incidents;
  (c) procedures in the event of partial unavailability.

(2) Future power stations - design aid (risk allocation).

Appendix 2

Criteria Adopted by the NRC*

The risk of immediate death as a consequence of an accident among the population located within a radius of 1.6 km from a power station must not exceed 1/1000th of the sum of the risks of accidental death of the average American.

The risk of cancer caused by an accident among the population located within a radius of 80 km from a power station must not exceed 1/1000th of the total risk of death by cancer of the American population.

To come below the limits specified above, industry must undertake any modification costing less than US$1000 per man X rem saved.

---

* The Nuclear Regulatory Commission.

Appendix 3

Co-operation Between the Nuclear Industry and the Oil Industry

EURODATA

Eurodata (European Association of Holders of Reliability Data Banks) was created in 1973. Its tasks were to resolve the problems of setting up and administering data banks, and to restore a common language providing for exchange of data between the various members of the Association.

The members from the nuclear industry are Ispra Research Centre (from the EEC), EDF, and UKAEA (the United Kingdom Atomic Energy Authority). The oil industry members are AGIP and BP. Eurodata is an institution financially supported by the ECC.

RISK ANALYSIS WORKING PARTY IN FRANCE

Under the auspices of the Chambre Syndicale de la Recherche et de la Production de Pétrole et de Gaz Naturel, French oil companies have set up a working group on risk analysis, aimed at providing an exchange of knowledge between the various members. Nuclear industry members are the civilian CEA, the military CEA, and EDF. Members from the oil industry are IFP, SNEA(P), and TOTAL-CFP.

Appendix 4

Co-operation Between Oil Companies on Systems Safety Research

CFP-IFP-SNEA(P) have undertaken to carry out a research programme at their own cost over the period 1982-84, to cover all techniques in systems safety and reliability. Work undertaken under this research programme comprises the following information.

THE COLLECTION OF RELIABILITY DATA

(1) Setting up a preliminary collection of data already in existence (a course already followed by EDF).

(2) Producing a compilation of experimental data on drilling and work-over equipment and operations (use of the DELPHI method).

(3) Producing a compilation of data on human reliability, from nuclear data and the analysis of experimental data.

THE ADAPTATION OF RELIABILITY ANALYSIS METHODS

(1) Analysis of existing methods.

(2) Production of case studies (including risk analysis on blow-out in work-over, in an endeavour to select the best methods).

(3) Development of availability and productivity design codes.

CONSEQUENCE ANALYSIS

(1) Development of a method for calculating the effects of an explosion in a confined atmosphere or otherwise.

(2) Investigation of the best models for gas cloud dispersion and fire radiation.

THE ACCEPTABLE RISK CRITERION

Analysis of criteria used in the various industries (chemical, nuclear, aeronautical) and various countries (France, Great Britain, United States).

Appendix 5

Drilling Rig Robotisation and Automation

Table A.5.1. An example of planning development

Axis	Item	1983	1984	1985	1986	1987
RIG OPERATION/ROTATION	BIL.K (BLM)					
	- Screwing/unscrewing mechanism					
	- Platform centring drillstring					
	- Semi-automatic slips					
	- Joint and lubrication positioning tool					
	- Automatic screwing/ unscrewing device					
	- Semi-automatic aid for the derrick man					
	- Floor racking system					
	- Automatic height adjustments					
	- BOP maintenance					
	- Semi-automatic drilling					
RIG OPERATION /ROTATION	- Hydraulic power swivel					
	- Electric power swivel					
	- Automatic power swivel					
	- Platform robotisation					
PLATFORM CONNECTION RACK	- Platform intake arm					
	- Intake mechanism					
	- Auxiliary maintenance and unit automisation					
CASING	- Semi-automatic tubing mechanism					

Study stage	Design stage	Trial stage
Specifications	Prototype construction	Industrial production
Basic design	Works test	Marketing/selling
Detailed design	Site test	

THE NATIONAL SUPPLY PROJECT

National Supply are currently carrying out site tests on an automatic system which could potentially make major changes in offshore operations. With no winch, rotary table, travelling block, hook or swivel, it eliminates the need for personnel to lower casings and, in fact, the majority of floor personnel. A single operator can carry out the drilling, controlling the rig from a remote control panel inside an enclosed cabin. The automatic rig can also be used as a slant rig, extending the platform's range, particularly where reserves are not at great depth.

National Supply have begun field tests on their automatic hydraulic rig designed for offshore duty. The rig is known as the National A 10-32 and uses an efficient microprocessor system for controlling and directing operations. The design drilling depth is 20 000 ft. (6000 m).

The A 10-32 project includes a power swivel moved up or down by four hydraulic lifting units, each ten inches in diameter. This principle eliminates the need for winches, rotary tables, travelling blocks, hooks, swivels, kellies, rat holes, mouse holes and virtually all minor handling items found on a conventional rig. It also virtually eliminates the need for floor personnel, the derrick man and personnel required to lower the casings.

All tubular equipment (casings, pipes and tubing) is manipulated by means of a boom which takes the pipe from the storage yard for use in the well, placing it in a vertical position, precision centred, above the well. Once the single or treble has been connected to the power swivel, addition to the rest of the drill string in the well is carried out. Rotation or circulation can be started in a matter of seconds.

The advantages of these possibilities are self-evident. If constriction is encountered when a casing is being lowered, the pumps can be re-started in a matter of seconds, and circulation can be resumed. Drilling or coring involves merely a brief stop near the bottom for stabbing. There is no string manoeuvring.

Robert E. Harris, President of National's Drilling Equipment Division, has declared the preliminary tests on the rig (completed in mid June 1982 at the company's base in Garland, Texas) to have been successful. During the test, the A 10-32 simulated drilling operations in a 4000 ft. well (1200 m). Field tests already started are scheduled to take six to twelve months. "These site tests represent an important part of the research efforts in the development of an automatic drilling rig", said R.E. Harris. "As soon as we are able to evaluate the results of the tests and carry out the additional modifications required, we hope we shall be able to offer a commercial version of the A 10-32. The importance of the A 10-32 as an automatic drilling rig rests in the fact that it replaces manual functions such as drill pipe handling and make-up, which have escaped automation for the 80 year history of rotary drilling", added R.E. Harris.

There have been no innovations in automatic rig design since the research into automatic rigs undertaken in 1960. The A 10-32, however, prefigures microprocessor controlled systems, hydraulically more powerful than its predecessors. Automatic rig prototypes have also all been onshore rigs, this being the first offshore system design.

The various advantages stated are as follows.

(a) As one man directs the rig from the remote control panel in an enclosed cabin, the rig may continue drilling in weather conditions that would stop a conventional offshore installation.

(b) Automatic manipulation of the drill pipes eliminates the majority of site personnel, thus making the rig safer, whether onshore or offshore.

(c) The automatic rig is more suitable for slant hole drilling than a conventional rig. This could extend the use of the rig and could be of considerable value in the case of multiple development wells from an offshore platform.
(d) The rig can resume rotating and/or circulating the mud within a matter of seconds and at any time in the operations.

There are further advantages in the case of offshore work in particular, aimed at reducing risks and accidents on the rig: living space requirements can be reduced. However, the increase in productivity is the major advantage. And an increase in productivity is invariably followed by healthy expansion and a greater demand for activity. There are notable examples of this in the oil industry. When automation was introduced in refineries, this involved elimination of the work carried out by distillation column workers, pump operators and gaugers. But at the same time, automation extended the possibilities of the industry, and in particular the manufacture of chemicals, plastics, etc., at least doubling manpower requirements.

It seems likely that automatic rigs will, in the same way, eliminate the majority of the manual and dangerous jobs on rigs. Similarly, the increase in productivity will cause more wells to be drilled, will increase the discovery of hydrocarbons and even the extraction of other minerals. This will require a larger and even more sophisticated labour force.

This study by National Supply represents an interesting approach to the problem. It should be noted that the National Supply project was totally abandoned for economic reasons (drilling crisis in the USA), although the objectives still remain a priority for the industry as a whole.

Appendix 6

Blow-out Risks Whilst Drilling - General Procedure

PREVENTIVE PHASE

Assessment of Blow-out Risk

Definition

A show of gas, oil or water is a penetration into the well of an effluent contained in a porous rock encountered in drilling (or workover on a well already drilled).

Causes

The effluent contained in a porous rock at the bottom of the well can only come out of that rock if the pressure in the well is lower than the pressure in the rock. Anything contributing to reduce the pressure at the bottom of the well may thus be the cause of a show if the pressure passes below a threshold, being the (pore) pressure of the rock. Wells are generally drilled with a filling of 'mud' to ensure that the pressure at the bottom is equivalent to that of the hydrostatic column constituted by this mud, of a known and controlled density.

The principal causes are as follows:

(a) excessively low mud density (ignorance of pressure systems, or negligence);
(b) fall in the mud level, hence an excessively low hydrostatic head (loss of mud in fractured rock, for example, negligence in pipe filling in tripping);
(c) trip swabbing, causing negative pressure at the bottom.

Indicators

The most reliable indicators, being the most easily identifiable, are:

(a) increase in the speed of advance (this being directly linked to the difference in pressure between the hole and the pores in the rock);
(b) loss of circulation, i.e. the mud fails to return to the surface (and is hence being absorbed by a porous or fractured rock);
(c) mud cut with gas or water;
(d) increase in the return rate of the mud and the level of the mud tanks (the mud comes out of the well, 'pushed' by the foreign effluent);
(e) lower filling during tripping than the volume required (hence another fluid is present in the well to make up the volume).

Preventive Procedures

Systems Inventory (Fig.A.6.1.)

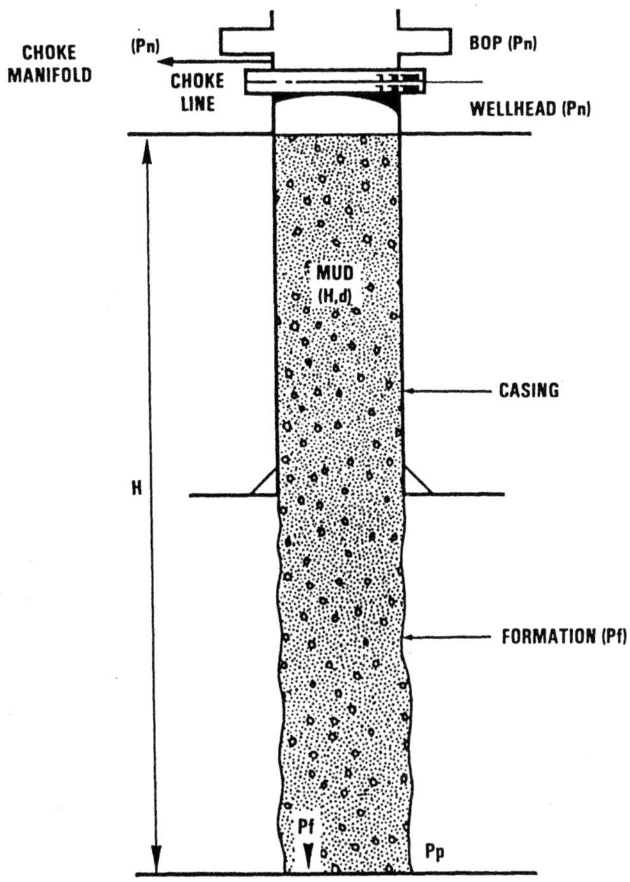

Fig.A.6.1.

The system used to control show are:

(a) formation, described by its pore pressure Pp, its fracturation limit pressure Pf;
(b) drilling mud (or completion fluid) described by its hydrostatic column, i.e. pressure Ph at the bottom (density x height);
(c) the last casing put into position - the level of its shoe, its mechanical characteristics (with particular reference to bursting strength);
(d) the wellhead in place, design pressure;
(e) the safety system - BOPs (blow-out preventers) plus the choke manifold, their configuration and design pressure.

It is important to note that we could have listed the systems enabling the indicators to be detected. These are essentially sensors located at the critical points of the rig.

Responsibilities

A drilling crew organisational chart is given in Fig.A.6.2. This team essentially comprises the drilling contractor's personnel, headed by a representative of the client (the drilling supervisor or superintendent). A site manager (contractor's) is responsible for his crew. The (supervisory) client is responsible for the well. The contractor (site manager) has responsibility for the equipment and men.

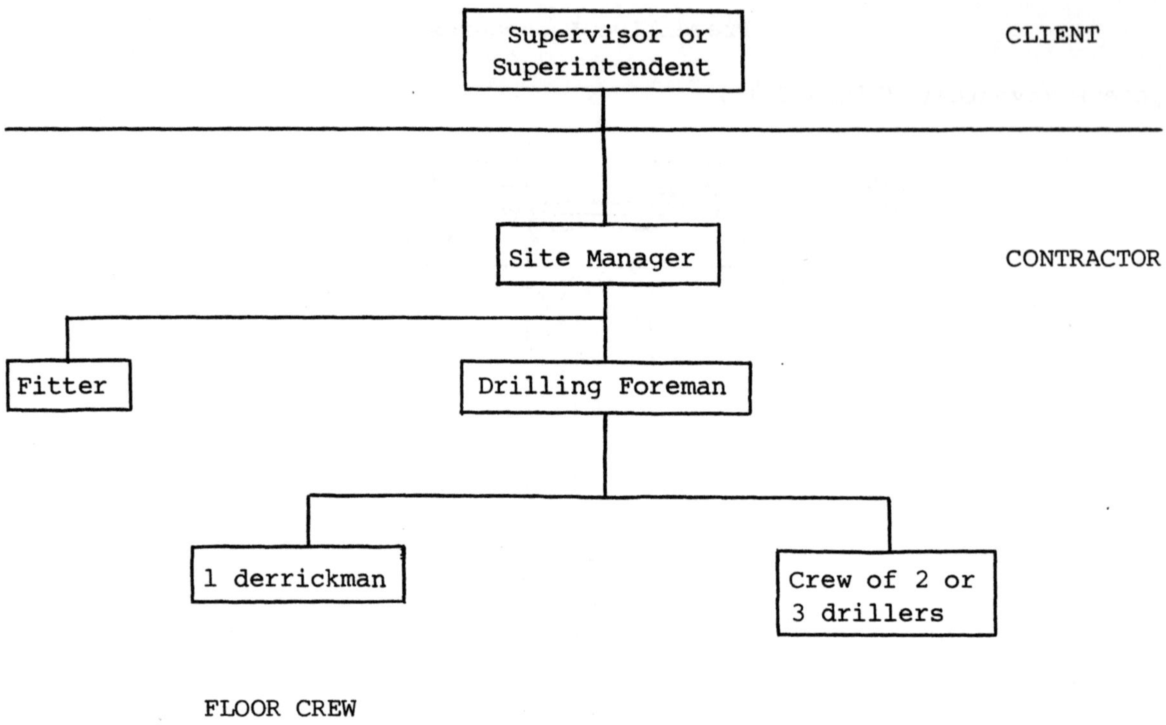

Fig.A.6.2.

On site, however, direct control of the rig is in the hands of a third person: the drilling foreman (at the brake on the drilling floor). It is therefore the drilling foreman who implements the initial procedures in the event of show, together with his floor crew.

Corrective Phase

(a) Level I. Return to normal procedure. (See Appendix 11: CFP procedure in the event of hydrocarbon show.)

(b) Level II. Escalation in the gravity of events - blow-out. A blow-out is the consequence of a show which it has not been possible to control using the methods specified in Appendix 11.

A review of the procedure to be followed to control a blow-out can be found in Appendix 15.

Appendix 7

Safety procedures in the Event of a Hydrocarbon Show While Drilling

EQUIPMENT AVAILABLE ON THE DRILLING RIG FOR PREVENTION, DETECTION AND SHOW CONTROL

Equipment Available for Detecting and Forecasting Shows

(a) Rate of advance recorder.
(b) Hydrometer (mud).
(c) D-exponent calculation.
(d) Chromatograph.
(e) Total gas detector with H_2S and CO_2 indicators.
(f) H_2S detector.
(g) Level gauge on all pits (fitted with alarm system).
(h) Totalising volume meter for the pits and trip tank.
(i) Weight indicator.
(j) Pressure gauges (a pressure gauge is fitted close to the choke manifold, connected to the riser).
(k) Flow return indicator in the return line, with alarm.

Equipment Available for Well Control

(a) BOP stack.
(b) 10 000 WP choke manifold (two variable chokes, one semi-automatic).
(c) Lower kelly cock (placed permanently under the square kelly).
(d) Upper kelly cock (placed permanently under the swivel).
(e) Drop-in type non-return valve fitting (situated in the drill string above the drill collars).
(f) Drill pipe float type non-return valve.
(g) Inside BOP Gray safety valve (on the floor, ready for quick fitting at the head of the drill string).
(h) Circulation head suitable for connection to the riser by means of 2 in. chicksans, 10 000 psi.
(i) Sufficient quantities of 2 in., 10 000 psi chicksan fittings.
(j) Swaco type degasser.
(k) Vertical degasser on possum belly tank connected to flare.
(l) Possum belly tank type inspection pit with visual gauge.
(m) 50 tonnes G grade cement.
(n) 60 tonnes barytes.
(o) 2 tonnes plugging agent.
(p) Heavy stand-by mud (1 tank).

ALERT CONDITIONS

Control procedures are to come into effect immediately on occurrence of any of the following.

(1) Increase in the volume of mud in the pits while drilling or well failing to fill normally while tripping.
(2) Total or partial loss.
(3) Increase of more than 50% in the rate of advance.
(4) Reduction in the pump pressure by more than 200 psi.
(5) Traces of gas on the chromatograph exceeding 5%.
(6) Mud emulsified.
(7) Appearance of oil on the mud.

ACTION TO BE TAKEN BY THE DRILLING FOREMAN

If alert conditions arise the shift drilling foreman is to advise the Total drilling superintendent and the contractor's site manager. At the same time, he is to initiate the control operations defined below.

While Drilling

(1) Stop drilling.
(2) Halve the pump rate.
(3) Release the square kelly so that the tool joint is 1 m above the rotary table.
(4) Stop the pump and check whether the well is flowing or losing:

(a) if neither flow nor loss applies, follow the instructions given by the site manager;
(b) if the well is losing, pump 100 litres of mud per minute into the annular space, awaiting instructions from the site manager;
(c) if the well is flowing, proceed as specified in (5), (6) and (7).

(5) Open the choke line valves.
(6) Close the annular type preventer at the recommended closing pressure (see pressure shown on the 'Standard Well Inspection Procedure').
(7) Gradually close the variable choke on the choke manifold, followed by the valve upstream from this choke. Every minute, record the pressure at the head of the pipes (P_t) and the pressure at the head of the annular space (P_a).

Note: do not exceed the maximum authorised pressure. Refer to the maximum allowable pressure shown in the Standard Well Inspection Procedure.

If this pressure is reached, circulate through the pipes at half the drilling rate, adjusting the choke in order not to exceed the maximum pressure allowable in the annular space.

While Tripping or Single Stabbing

(1) Immediately stop tripping.
(2) Screw in the kelly.
(3) Check whether the well is flowing or losing:

(a) if neither flowing nor losing, follow the instructions given by the site manager;
(b) if the well is losing, follow the instructions given by the site manager;

(c) if the well is flowing, proceed as specified in (4), (5) and (6).

(4) Follow instructions for (5) above.
(5) Follow instructions for (6) above.
(6) Follow instructions for (7) above.

Drill String Out of Hole

(1) Open the choke line valves.
(2) Close the blind rams.
(3) Gradually close the variable choke on the choke manifold. Every minute, record the pressure at the head of the casing (P_a).

Note: do not exceed the maximum authorised pressure. Refer to the maximum allowable pressure shown in the Standard Well Inspection Procedure.

If the pressure reaches or exceeds the maximum allowable pressure, bleed by discharging through the vertical degasser and then by the Swaco (mud) degasser or through the flare (oil or gas).

In the Well Control Manual, the upper line is termed the choke line and the lower line the kill line. Whenever possible, the upper line (the choke line) should be used for well control operations, the lower line (the kill line) being used only in cases of dire emergency.

ACTION TO BE FOLLOWED BY THE DRILLING FOREMAN AND CONTRACTOR'S SITE MANAGER

Increase in the Volume of Mud in the Pits

Immediately on becoming aware of the alert status, the drilling superintendent and site manager must go to the drilling foreman on the floor and ascertain the well condition. Assuming the drilling foreman to have duly performed the preliminary control operations, the drilling superintendent and the contractor's site manager are to proceed as below.

Show While Drilling

(1) Place in the drop-in valve.
(2) Bleed the drill pipe head pressure to ensure that the valve is not leaking. If not, pump gently to measure the P_t.
(3) Close the bottom kelly cock.
(4) Unscrew the square kelly above the bottom kelly cock and stow it in the rat hole.
(5) If the float valve is leaking, set the Gray type non-return valve above the kelly cock to the closed position.
(6) Screw the safety assembly and circulating head onto the drill string, connect it to the riser, and open the safety valve under the circulating head.
(7) Pressurise above the bottom kelly cock, to equalise pressures (only if the Gray type non-return valve is to be used).
(8) Open the bottom kelly cock and close the circulating head valve.
(9) Gently clear the drill string to bring out the slips.
(10) Close and lock the upper pipe rams.
(11) Open the annular type preventer and check the upper pipe rams for leak-tightness.
(12) Slowly lower the drill string to lay the tool joint on the upper pipe rams. Re-check these closures for leak-tightness. If leaking, close the annular type preventer.

(13) Open the circulating head valve.
(14) Control the well by the driller's method described in Appendix 9.

Note: after stabilising the well and before resuming drilling, a full test should be carried out on the preventers.

Trip Show

The well is flowing through the annular space: this show is the result of inadequate filling of the well during pull-out and/or swabbing with a mud of a density too close to the reservoir pressure gradient. Bleed the pressure at the choke manifold to measure the flow.

(a) Flow less than 10 gallons/minute (40 litres/minute)

(1) Close the variable choke at the choke manifold.
(2) Bleed the pressure inside the drill pipes to ensure that the float valve is leak-tight.
(3) Remove the square kelly.
(4) Open the annular type preventer.
(5) Lower the drill string as far as possible.
(6) Close the annular preventer.
(7) Connect the circulating head and fit to the riser.
(8) Slowly clear the drill string to bring out the slips.
(9) Close and lock the upper pipe rams.
(10) Open the annular type preventer and ensure that the upper pipe rams are leak-tight, observing the well.
(11) Slowly lower the drill string to place the string on the upper pipe rams. Re-check these closures for leak-tightness. If leaking, close the annular type preventer.
(12) Calculate the mud density required to increase the bottom pressure by 200 psi.
(13) Weight the surface mud to this value.
(14) Open the circulating head valve.
(15) Circulate the volume of mud between the tool and the aperture using the driller's method.
(16) If, after this circulation, the well is flowing less than 10 gallons/minute, resume descent to the bottom and homogenise the mud.
(17) If, after this circulation, the well is yielding more than 10 gallons/minute, squeeze the volume between the bottom and the tool.

(b) Flow above 10 gallons/minute

(1) Close the variable choke of the choke manifold.
(2) Bleed the pressure inside the drill pipes to ensure that the float valve is leak-tight.
(3) Close the bottom kelly cock.
(4) Unscrew the square kelly above the bottom kelly cock and stow it in the rat hole.
(5) If the float valve is leaking, set the Gray type non-return valve above the kelly cock to the closed position.
(6) Screw the circulating head and connect it to the riser.
(7) Pressurise above the bottom kelly cock, to equalise the pressures (only if the Gray type non-return valve is to be used).
(8) Open the bottom kelly cock.
(9) Gently clear the drill string to bring out the slips.
(10) Close and lock the upper pipe rams.

(11) Open the annular type preventer and check the upper pipe rams for leak-tightness.
(12) Slowly lower the drill string to lay the tool joint on the upper pipe rams. Re-check these closures for leak-tightness. If leaking, close the annular type preventer.
(13) Calculate the mud density required to increase the bottom pressure by 200 psi.
(14) Weight the surface mud to this value.
(15) Circulate the volume of mud between the tool and the aperture using the driller's method.
(16) If, after this circulation, the well is yielding less than 10 gallons/minute, resume descent to the bottom and homogenise the mud.
(17) If, after this circulation, the well is yielding more than 10 gallons/minute, squeeze the volume between the bottom and the tool.

After stabilising and before resuming drilling, carry out a full test on the preventers.

Show with the Drill String Outside the Hole

This show is the result of inadequate well filling during pull-out and/or swabbing due to a mud of a density too close to the reservoir pressure gradient. There are two possibilities:

(1) low pressure in the casing or any other indication of slight show; and
(2) rapid pressure rise or any other indication of heavy show.

(a) Low pressure in the casing or any other indication of slight show

(1) Check, via the choke manifold, that the flow is stable and does not exceed 10 gallons/minute. Close the variable choke.
(2) As quickly as possible, lower the pipes fitted with safety valves at the ends.
(3) When the pipes are close to the annular type preventer, open the blind rams.
(4) Resume descent. A gain of 10 bbls in a hole diameter 12 in. or 4 bbls in a hole diameter 8½ in. is acceptable. Only fill the pipes if this is necessary to avoid the collapse effect.
(5) As soon as the bottom is reached, or at the limit of allowable gain, close the annular preventer and fill the pipes.
(6) Connect the circulating head fitted to the riser.
(7) Slowly clear the drill string to bring out the slips.
(8) Close and lock the upper pipe rams.
(9) Open the annular type preventer and check the closures on the pipes for leak-tightness.
(10) Slowly lower the drill string to place the string on the upper pipe rams. When the closure is in place, re-check these closures on the pipes for leak-tightness. If leaking, close the annular type preventer.
(11) Calculate the mud density required to increase the bottom pressure by 200 psi.
(12) Weight the surface mud to this value.
(13) Circulate the volume of mud between the tool and the aperture using the driller's method.
(14) If, after this circulation, the well is yielding less than 10 gallons/minute, resume descent to the bottom and homogenise the mud.

(15) If, after this circulation, the well is yielding more than 10 gallons/minute, squeeze the volume between the bottom and the tool.

(b) <u>Rapid rise in pressure or any other indication of show of more than 10 gallons/minute</u>

(1) Calculate the mud density required to balance the well plus a bottom pressure of 200 psi.
(2) Adjust the density of the spare mud to the required value, taking an adequate safety margin (volume).
(3) Squeeze the volume of mud necessary to kill the well through the choke line.
(4) Observe the stability of the well through the choke manifold. If the well is not killed, repeat the operation.
(5) When the well is killed,
(6) Open the blind rams. Close the choke manifold. There are two possibilities:
 (i) the well is killed and not losing returns;
 (ii) the well is killed and losing returns.

(i) <u>The well is killed and not losing returns</u>
(1) Start lowering the tool with a safety valve.
(2) Whilst tripping, check the stability of the well and stop the trip if the mud return is greater than the drill string displacement. A gain of 10 bbls is acceptable in a hole diameter 12¼ in., and 5 bbls in a hole diameter 8½ in.
(3) As soon as the bottom is reached, or at the limit of allowable gain:
(4) Close the annular preventer (if the well is flowing).
(5) Fit the circulating head connected to the riser.
(6) Slowly clear the drill string to bring out the slips.
(7) Close and lock the upper pipe rams.
(8) Open the annular type preventer and check the closures on the pipes for leak-tightness (if the annular type preventer has been closed).
(9) Slowly lower the drill string to place on the upper pipe rams. Re-check the closures for leak-tightness. If leaking, close the annular preventer.
(10) With the driller's method, circulate the volume of mud contained between the tool and the aperture.
(11) If, after this circulation, the well is discharging less than 10 gallons/minute, resume descent to the bottom and homogenise the mud.
(12) If, after this circulation, the well is discharging more than 10 gallons/minute, squeeze the volume between the bottom and the tool

(ii) <u>The well is killed and losing returns</u>
(1) Pump mud into the annular space to keep the well filled with a maximum of 20 m^3/hr.
(2) Lower the drill string and proceed as indicated below ('Show while drilling').

Total or Partial Mud Loss

Immediately upon becoming aware of the alert status, the contractor's site manager must go to the drilling foreman on the floor and ascertain the well conditions. Assuming the drilling foreman to have duly performed the preliminary control operations, the drilling superintendent and the contractor's site manager are to proceed as follows.

Show While Drilling

There are two possibilities, depending on the extent of loss.

(1) <u>Partial loss $< 3\ m^3/hr$</u>

(a) Circulate with the mud used, checking loss.

(b) Reduce loss by:
- (i) reducing the density of the mud;
- (ii) injecting one or more plugs of plugging agent;
- (iii) adding plugging agents to the mud.

(2) <u>Partial loss $> 3\ m^3/hr$ or total loss</u>

(a) Pump into the annular space to keep the well full.

(b) If the well can no longer be kept full, permanently pump at the maximum rate of $20\ m^3/hr$.

(c) Raise the tool to the shoe or 150 m above the bottom.

(d) Set the square kelly in place, and resume circulation.

(e) Reduce loss by:
- (i) reducing the density of the mud;
- (ii) injecting one or more plugs of plugging agent;
- (iii) injecting a plug of cement or cement gel.

Trip Losses

There are two possibilities, depending on the extent of the loss.

(a) <u>Partial loss $< 3\ m^3/hr$</u>

(a) Pump into the annular space to keep the well full.

(b) Return the drill string to the bottom.

(c) Fit the square kelly.

(d) Resume rotation and circulation at a reduced rate.

(e) Reduce losses by:
- (i) reducing the density of the mud;
- (ii) injecting one or more plugs of plugging agent;
- (iii) adding plugging agents to the mud.

(b) <u>Partial loss $> 3\ m^3/hr$ or total loss</u>

(a) Pump into the annular space to keep the well full.

(b) If the well can no longer be kept full, permanently pump at a maximum rate of $20\ m^3/hr$.

(c) Raise the tool to the shoe or 150 m above the bottom.

(d) Set the square kelly in place, and resume rotation and circulation at a reduced rate.

(e) Reduce loss by:
- (i) reducing the density of the mud;
- (ii) injecting one or more plugs of plugging agent;
- (iii) injecting a plug of cement or cement gel.

Increase of More than 50% in the Rate of Advance

Immediately on becoming aware of the alert status, the contractor's site manager must go to the drilling foreman on the floor and ascertain the well conditions. Assuming the drilling foreman to have duly performed the preliminary control operations, the drilling superintendent and the contractor's site manager are to proceed as follows.

Well No Longer Flowing

Resume drilling, checking frequently that the well is not flowing.

Well Flowing

Proceed as stated in the third section (Instructions for the drilling foreman) from 'While drilling', §4.

Well Losing Returns

Proceed as indicated above in 'Show while drilling', §2, partial or total loss.

Reduction of More than 200 psi in the Pump Pressure, Traces of Gas on the Chromatograph, Exceeding 5%, Mud Emulsified, or Appearance of Oil on the Mud

Immediately upon becoming aware of the alert status, the contractor's site manager must go to the drilling foreman on the floor and ascertain the well conditions. Assuming the drilling foreman to have duly performed the preliminary control operations, the drilling superintendent and the contractor's site manager are to proceed as follows.

Well Not Flowing

Resume rotation and circulation. After one circulation cycle, resume drilling if no anomalies observed.

Well Flowing

Open the choke line valves.

Well Losing Returns

Proceed as stated above 'Total or partial mud loss' - 'Show while drilling'.

RULES FOR ABANDONMENT OF THE WELL IN THE EVENT OF BLOW-OUT RISK

Leak of Last Rams on Drill Pipes

(1) Close the shear rams to cut off the pipe.
(2) Using the kill line, squeeze a quantity of cement capable of filling 1000 m of the hole. Flush the cement with sufficient mud to clean the kill line and wellhead.
(3) Close the well for a minimum period of eight hours.
(4) Whilst waiting for the cement to set, note pressure behaviour.

<u>Note</u>: all persons whose presence is not necessary for the hole control operations are to be evacuated.

Blind Rams Not Leak-tight

(1) Close the safety valves on the choke and kill lines.
(2) Raise the string of drill pipes cut-off above the BOP.

Note: plugging agents such as leather or rubber are effective ways of sealing the rams if added to the mud or cement: this should be borne in mind whenever critical rams are leaking.

Appendix 8

Pressure Tests on Formations in the Immediate Vicinity of Riser Shoes

GENERAL

The suitability of carrying out a pressure test on the formations after using a protection casing should be investigated when preparing the drilling programme, and the opinion of the Operations Department should be sought whenever operational sectors feel such to be useful.

Any pressure test on formations under the shoe should, therefore, be mentioned in the corresponding phase of the drilling programme, and in such cases it should be stated whether the formation is to be tested at a fixed pressure regarded as sufficient in the light of expected problems, or to the leak-off pressure.

OBJECT OF TEST

The object of this test is to prove the strength of the ground around the shoe of the casing at a certain preset pressure value or leak-off pressure, and verify the quality of cementing.

Pressure Not to be Exceeded at the Surface During the Performance of the Test

The maximum pressure which must not be exceeded at the surface (annular space and inside pipes) is the working pressure of the plugs or the leak test pressure specified for the casing, whichever is the lower.

Calculation of the Leak Test Pressure Possible on a Casing as a Function of the State of the Well

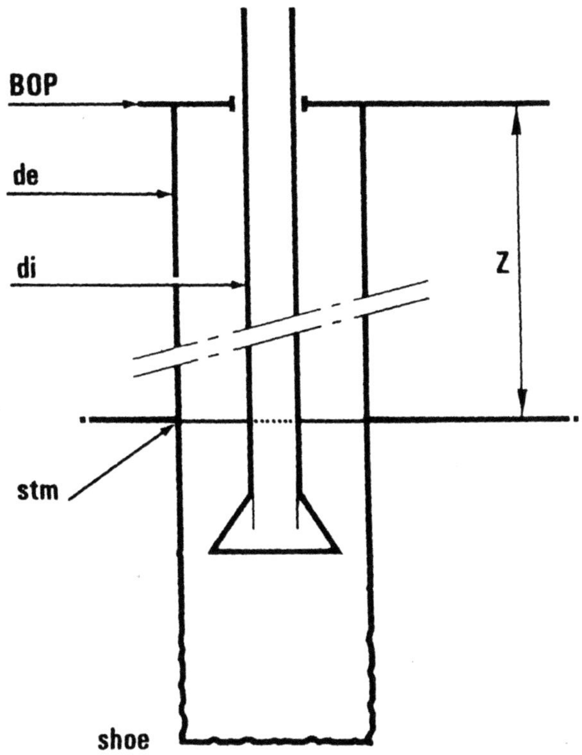

stm = Section of casing most submitted to burot

z = Vertical depth of Stm section

di = Mud density inside casing

de = Mud density outside casing

P_E = Burst value at botton of section Stm most submitted to burst (kg/cm²)

P_{Max} = Maximum test pressure (kg/cm²)

$$P_{Max} = 0{,}80\ PE \cdot \frac{(di \cdot de) \times Z}{10} \quad \text{in kg/cm}^2$$

Fig.A.8.1.

METHOD OF PERFORMING THE PRESSURE TEST UNDER THE SHOE

After performing the leak tests on the control items (plugs etc.) and casing, the pressure test under the shoe is to be performed as follows:

(1) Redrill the valve, cement and shoe.
(2) Drill ten metres of formation (if there is any doubt as to the homogeneity of the zone covered, drill to a maximum of 50 m).
(3) Circulate until the density of the mud is homogenised.
(4) Raise the tool to the shoe.
(5) Close the annular space.
(6) Inject the mud through the pipes with the cementing unit at a constant rate between 50 and 100 litres per minute.

(7) Plot the pressure rise vs the pumped volume on a graph (1 point every 50 or 100 litres measured at the cementing unit tanks). In parallel, record this pressure rise.

(8) Stop the pump when:

(a) the pressure at the surface reaches a preset value regarded as sufficient to cope with the problems expected during the drilling phase;

(b) two consecutive points deviate substantially from the mean pressure rise straight line passing through the other points. The divergence point (see example) marks the value of the pressure at the head from which mud begins to be injected into the formation (the leak-off) point.

(9) After stopping the pump, maintain the pressure for ten minutes and record the pressure drop minute by minute.

(10) Release the pressure and measure the volume of mud recovered. Compare it with the volume recovered in the casing leak test.

CALCULATION OF ALLOWABLE EQUIVALENT DENSITY ON FORMATIONS IN THE VICINITY OF THE SHOE

Where:

P_{adm} = the surface pressure corresponding to the leak-off point, in kg/cm^2

Z_s = vertical elevation of the shoe, in metres

P_{fond} = bottom pressure corresponding to elevation Z_s, in kg/cm^2

then:

$$P_{fond} = P_{adm} + \frac{di \times Z_s}{10} \quad \text{in kg/cm}^2$$

hence the value of the equivalent leak-off density is:

$$d_{eq} = \frac{P_{fond} \times 10}{Z_s}$$

Assuming a safety margin of 0.05, the maximum allowable equivalent density on the formation in the vicinity of the shoe will be:

$$d_{eqv_{max}} = d_{eq} - 0.05$$

The equivalent density 'd_{eqv}' is, of course, the value of the pressure at a given vertical elevation, related to that elevation. It is expressed in kg/cm^2 per 10 m vertical depth, for assimilation with a density, if the column of mud is immobile and the level at the aperture 'd_{eqv}' is equal to the density of the mud.

In drilling, d_{eqv} is equal to:

$$d_{eqv} = da + \frac{10 \times P_a}{Z}$$

where:

da = density of mud in the annular space

P_a = pressure drops in the annular space for a given flow rate in kg/cm^2

Z = vertical elevation in question, in metres.

In tripping, as a result of swabbing, d_{eqv} is greater than the density of the mud on descent, and less on pull-out.

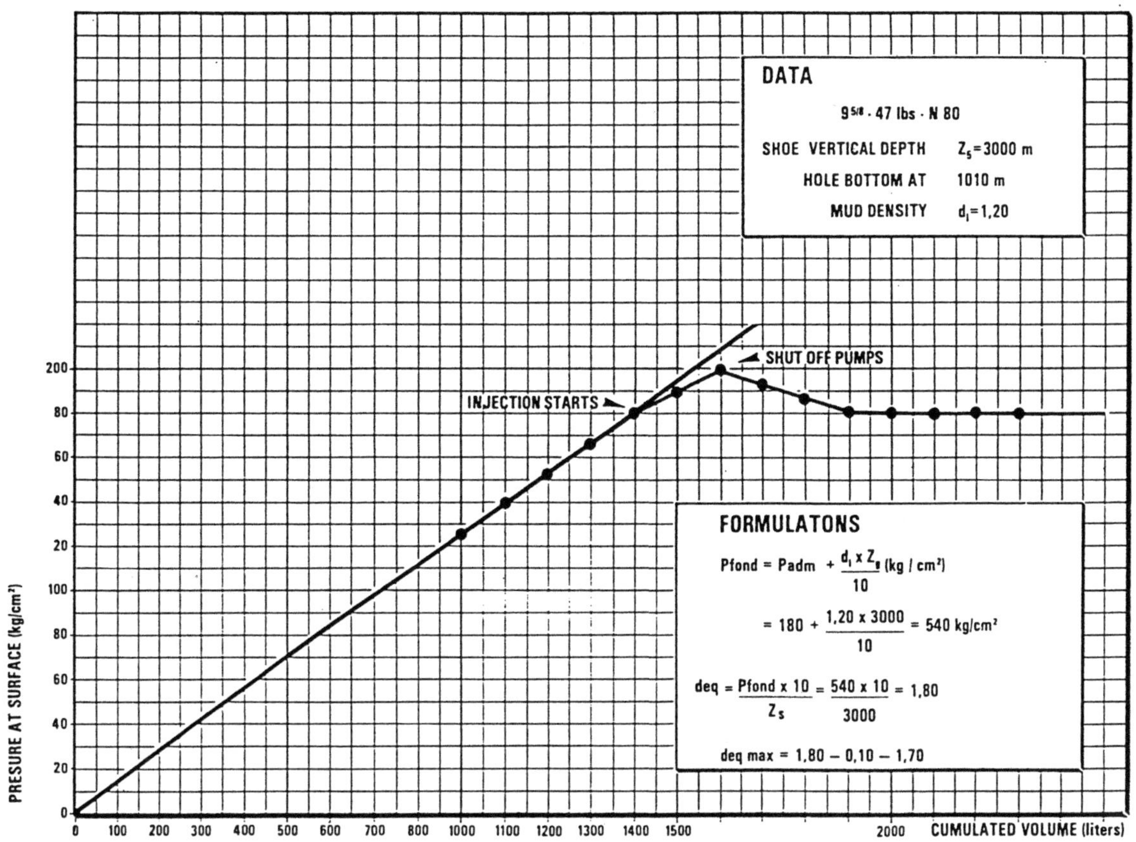

Fig.A.8.2.

Appendix 9

Well Control by the 'Driller's Method'

EVACUATION OF A FLUID HAVING PENETRATED THE WELL

(1) Open the valve upstream of the variable choke.
(2) Resume circulation at the selected reduced rate, and keep this rate constant.
(3) At the same time, open the variable choke to maintain a constant pressure on the annular space equal to the annular closing pressure plus a safety factor S.
(4) When the annular pressure has stabilised, note the well pipe head pressure and keep it constant by actuating the adjustable choke until a full circulation cycle has been completed.

CALCULATION OF THE DENSITY REQUIRED FOR WELL CONTROL

(1) Pipe head closing pressure: Pt_1 in kg/cm^2.

Safety factor S (generally 5 kg/cm^2 per 1000 m).

Depth of well: z, in metres.

Density if circulating mud: initial density (dl) - density required (dr).

$$dr = dl + 10 \frac{Pt_1}{z} + 0.05$$

(2) Production of a volume of mud equal to the volume in circulation at the calculated density dr.

PLACEMENT OF MUD OF THE REQUIRED DENSITY dr

(1) Inject the mud, density d_r, keeping the annular pressure constant at (Pt_1 + S) by actuating the variable choke.

(2) When the internal volume of the pipes has been pumped, note the pipe head pressure and keep constant over the end of the circulation cycle.

Appendix 10

'BOP' Tests

OPERATING TESTS

A pure operating test is to be carried out on each trip.

PRESSURE TESTS

(1) Working pressure

All equipment is to be tested at the working pressure (10 000 psi) or 5000 psi) at the start of the contract.

(2) To suit the well condition

Every two weeks, after each casing and before entering reservoirs, control devices are to be tested at the maximum pressure expected to arise in the relevant phase (see drilling schedule).

CASING TESTS

Casings are to be tested at the end of flush surge pressure. Their annular seals are to be tested at the same time as the BOP stack, with care (because of the risk of collapse).

All tests above are to be mentioned in the daily reports.

Appendix 11

Shift Manager's Instructions (in the event of well shows observed at the surface) - Drilling on Land or on a Fixed Platform

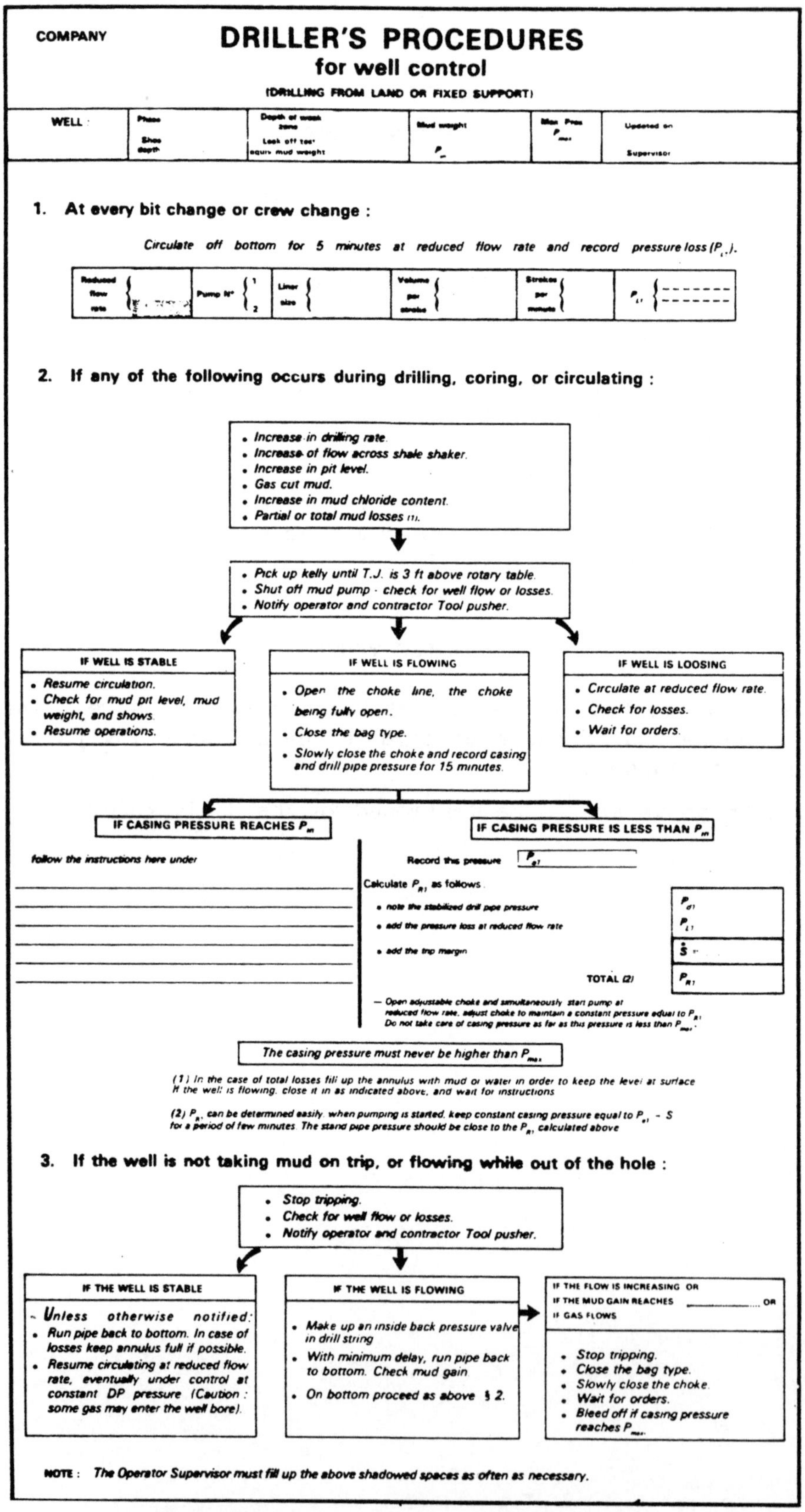

COMPANY

DRILLER'S PROCEDURES
for well control
(DRILLING FROM LAND OR FIXED SUPPORT)

WELL	Phase / Shoe depth	Depth of weak zone / Leak off test equiv. mud weight	Mud weight P_m	Max. Pres. P_{max}	Updated on / Supervisor

1. At every bit change or crew change :

Circulate off bottom for 5 minutes at reduced flow rate and record pressure loss (P_L).

Reduced flow rate	Pump N° 1 / 2	Liner size	Volume per stroke	Strokes per minute	P_L

2. If any of the following occurs during drilling, coring, or circulating :

- *Increase in drilling rate.*
- *Increase of flow across shale shaker.*
- *Increase in pit level.*
- *Gas cut mud.*
- *Increase in mud chloride content.*
- *Partial or total mud losses (1).*

↓

- *Pick up kelly until T.J. is 3 ft above rotary table.*
- *Shut off mud pump - check for well flow or losses.*
- *Notify operator and contractor Tool pusher.*

IF WELL IS STABLE
- *Resume circulation.*
- *Check for mud pit level, mud weight, and shows.*
- *Resume operations.*

IF WELL IS FLOWING
- *Open the choke line, the choke being fully open.*
- *Close the bag type.*
- *Slowly close the choke and record casing and drill pipe pressure for 15 minutes.*

IF WELL IS LOOSING
- *Circulate at reduced flow rate.*
- *Check for losses.*
- *Wait for orders.*

IF CASING PRESSURE REACHES P_m

follow the instructions here under

IF CASING PRESSURE IS LESS THAN P_m

Record this pressure P_{a1}

Calculate P_{R1} as follows

- *note the stabilized drill pipe pressure* P_{d1}
- *add the pressure loss at reduced flow rate* P_{L1}
- *add the trip margin* S

TOTAL (2) P_{R1}

— *Open adjustable choke and simultaneously start pump at reduced flow rate, adjust choke to maintain a constant pressure equal to P_{R1}. Do not take care of casing pressure as far as this pressure is less than P_{max}.*

The casing pressure must never be higher than P_{max}

(1) In the case of total losses fill up the annulus with mud or water in order to keep the level at surface. If the well is flowing, close it in as indicated above, and wait for instructions.

(2) P_{R1} can be determined easily when pumping is started, keep constant casing pressure equal to P_{a1} - S for a period of few minutes. The stand pipe pressure should be close to the P_{R1} calculated above.

3. If the well is not taking mud on trip, or flowing while out of the hole :

- *Stop tripping.*
- *Check for well flow or losses.*
- *Notify operator and contractor Tool pusher.*

IF THE WELL IS STABLE

- *Unless otherwise notified:*
- *Run pipe back to bottom. In case of losses keep annulus full if possible.*
- *Resume circulating at reduced flow rate, eventually under control at constant DP pressure (Caution : some gas may enter the well bore).*

IF THE WELL IS FLOWING
- *Make up an inside back pressure valve in drill string*
- *With minimum delay, run pipe back to bottom. Check mud gain.*
- *On bottom proceed as above § 2.*

→ IF THE FLOW IS INCREASING OR IF THE MUD GAIN REACHES .... OR IF GAS FLOWS
- *Stop tripping.*
- *Close the bag type.*
- *Slowly close the choke.*
- *Wait for orders.*
- *Bleed off if casing pressure reaches P_{max}.*

NOTE : *The Operator Supervisor must fill up the above shadowed spaces as often as necessary.*

Fig.A.11.1. Driller's procedures for well control (drilling from land or fixed support).

KEY TO FIGURE A.11.1.

1 Site

2 At the start of each shift, or after changing tools: circulate for five minutes above the bottom at reduced rate Qr and note the pressure drop P_c 1.

3 If, whilst drilling, core sampling or bottom circulating, you experience:

4 - a marked increase in the rate of advance;
- an increase in the rate at the ditch;
- a gain in the mud in the pits;
- traces of gas or oil in the mud;
- an increase in chlorine in the mud;
- a loss of circulation (1).

5 You should:
- position to 1st T.J. 1 m above the table;
- stop circulation - observe the well;
- advise the client and contractor's senior representatives.

6 IF THE WELL IS STABLE:
- resume circulation and operations in progress;
- increase monitoring of levels, density and indicators.

7 IF THE WELL IS FLOWING:
- open the circuit under the choke with the choke fully open;
- close the annular type preventer;
- gradually close the choke, note P_a and P_t for 15 minutes.

8 If the annular pressure (P_a) reaches P_{adm}.

9 If the annular pressure (P_a) remains below P_{adm}.

10 Apply the following instructions.

11 Note the pressure.

12 Determine PR 1 as follows:
- note the head pressure of the pipes;
- add the pressure drop at reduced rate;
- add the safety factor.

13 Open the choke and at the same time resume circulation at constant reduced rate Or maintaining, by the choke, a constant return pressure equal to PR 1. Do not worry about P_a as long as its value remains below P_{max}.

14 Under no circumstances allow the annular pressure to exceed P_{max}.

15 (a) In the event of total loss, fill the annular space with mud or water to try and bring the level to the aperture. If the well is flowing, close as stated above, and await instructions.

16 (b) PR 1 may be established directly by keeping the annular pressure constant and equal to P_a 1 + S for a number of minutes on resuming circulation (compare with calculated PR 1).

17 If you experience anomalies in filling on tool pull-out or return on descent or well flow when the tool is above ground:

18 - stop the trip;
- observe the ditch and pipe heads;

- advise the client's and contractor's senior representative.

19 IF THE WELL IS NOT FLOWING
Unless instructed otherwise:
- return to the bottom;
- if losing, keep the well full if possible by pumping in the annular space
- resume circulation at reduced rate and possibly under pressure chokes at constant pipe head pressure (a plug of gas may be formed at the bottom).

20 IF THE WELL IS FLOWING
- screw down a valve at the drill pipe head.
- quickly return to the bottom, controlling the return volume;
- at the bottom, apply the instructions in 2 above.

21 IF THE FLOW RATE ACCELERATES OR IF GAS APPEARS IN THE RETURN LINE
- stop the trip;
- close the annular type preventer;
- gradually close the choke;
- await instructions;
- bleed if the annular pressure reaches P_{max}.

22 Note: the shaded areas in this table are to be completed and kept up to date by the person in charge of drilling.

23 IF THE WELL IS LOSING RETURNS
- circulate at reduced rate;
- check loss;
- await instructions.

Appendix 12

Blow-out Control Calculation Sheet

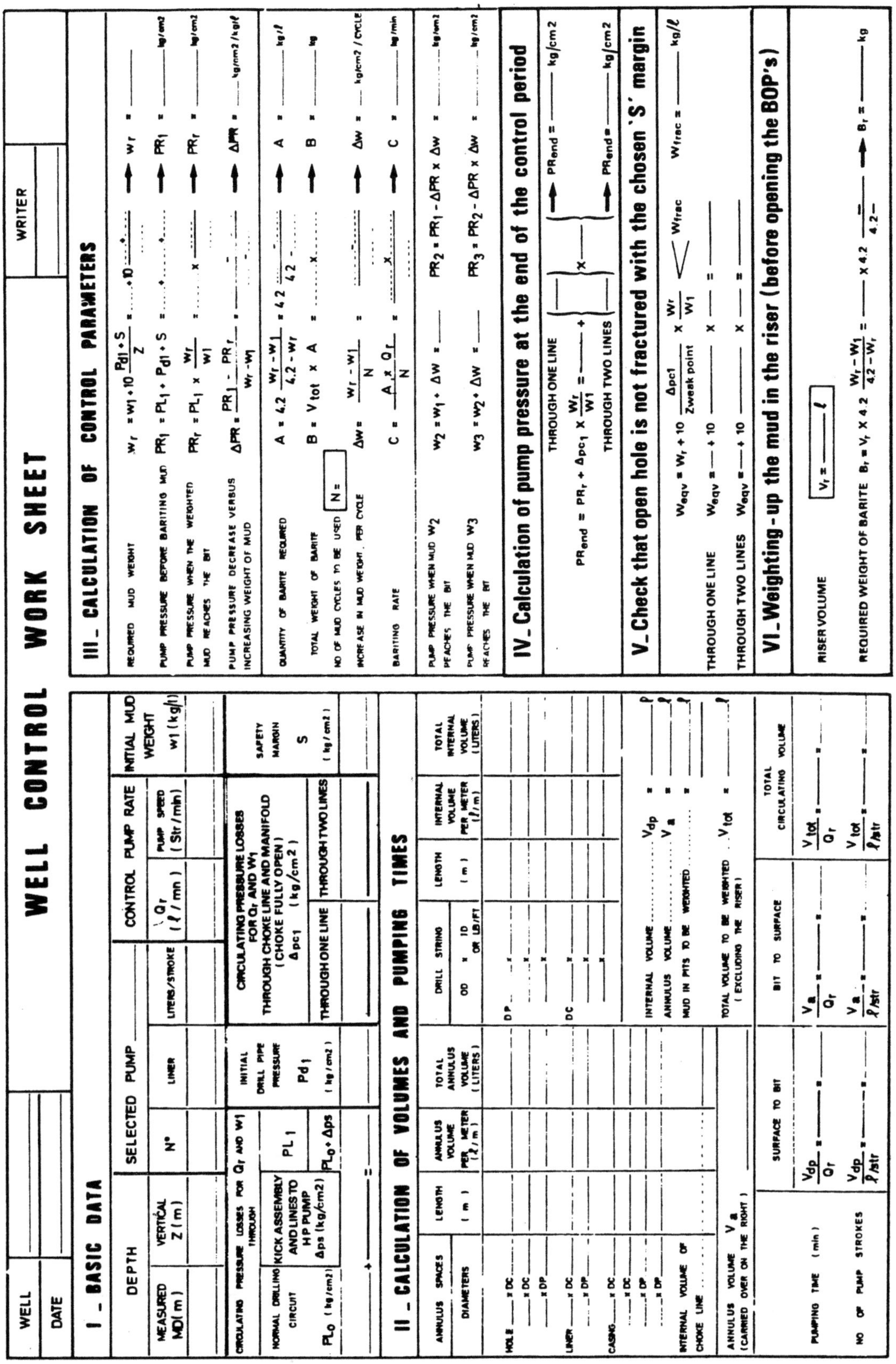

WELL
DATE
WELL CONTROL WORK SHEET
WRITER

I _ BASIC DATA

DEPTH: MEASURED MD(m) | VERTICAL Z(m)
SELECTED PUMP: N° | LINER | LITERS/STROKE
CONTROL PUMP RATE: Q_r (ℓ/mn) | PUMP SPEED (Str/min)
INITIAL MUD WEIGHT w_1 (kg/l)

CIRCULATING PRESSURE LOSSES FOR Q_r AND W_1 THROUGH: NORMAL DRILLING CIRCUIT PL_0 (kg/cm2) | KICK ASSEMBLY AND LINES TO HP PUMP Δps (kg/cm2) | PL_1 = $PL_0 + \Delta ps$
INITIAL DRILL PIPE PRESSURE Pd_1 (kg/cm2)
CIRCULATING PRESSURE LOSSES FOR Q_r AND W_1 THROUGH CHOKE LINE AND MANIFOLD (CHOKE FULLY OPEN) Δpc_1 (kg/cm2): THROUGH ONE LINE | THROUGH TWO LINES
SAFETY MARGIN S (kg/cm2)

II _ CALCULATION OF VOLUMES AND PUMPING TIMES

ANNULUS SPACES DIAMETERS | LENGTH (m) | ANNULUS VOLUME PER METER (ℓ/m) | TOTAL ANNULUS VOLUME (LITERS)
HOLE x DC, x DC, x DP
LINER x DC, x DP
CASING x DC, x DC, x DP, x DP
INTERNAL VOLUME OF CHOKE LINE
ANNULUS VOLUME V_a (CARRIED OVER ON THE RIGHT)

DRILL STRING OD x ID OR LB/FT | LENGTH (m) | INTERNAL VOLUME PER METER (ℓ/m) | TOTAL INTERNAL VOLUME (LITERS)
DP x, x, x
DC x, x, x
INTERNAL VOLUME V_{dp} = ℓ
ANNULUS VOLUME V_a = ℓ
MUD IN PITS TO BE WEIGHTED = ℓ
TOTAL VOLUME TO BE WEIGHTED (EXCLUDING THE RISER) V_{tot} = ℓ

	SURFACE TO BIT	BIT TO SURFACE	TOTAL CIRCULATING VOLUME
PUMPING TIME (min)	$\frac{V_{dp}}{Q_r}$ = =	$\frac{V_a}{Q_r}$ = =	$\frac{V_{tot}}{Q_r}$ = =
NO OF PUMP STROKES	$\frac{V_{dp}}{\ell/str}$ = =	$\frac{V_a}{\ell/str}$ = =	$\frac{V_{tot}}{\ell/str}$ = =

III _ CALCULATION OF CONTROL PARAMETERS

REQUIRED MUD WEIGHT: $w_r = w_1 + 10\frac{P_{d1}+S}{Z}$ = ... → w_r =
PUMP PRESSURE BEFORE BARITING MUD: $PR_1 = PL_1 + P_{d1} + S$ = ... → PR_1 = kg/cm2
PUMP PRESSURE WHEN THE WEIGHTED MUD REACHES THE BIT: $PR_r = PL_1 \times \frac{w_r}{w_1}$ = ... → PR_r = kg/cm2
PUMP PRESSURE DECREASE VERSUS INCREASING WEIGHT OF MUD: $\Delta PR = \frac{PR_1 - PR_r}{w_r - w_1}$ = ... → ΔPR = kg/cm2/kg/ℓ
QUANTITY OF BARITE REQUIRED: $A = 4.2\frac{w_r - w_1}{4.2 - w_r}$ = ... → A = kg/ℓ
TOTAL WEIGHT OF BARITE: $B = V_{tot} \times A$ = ... → B = kg
NO OF MUD CYCLES TO BE USED N =
INCREASE IN MUD WEIGHT, PER CYCLE: $\Delta w = \frac{w_r - w_1}{N}$ = ... → Δw = kg/cm2/CYCLE
BARITING RATE: $C = \frac{A \times Q_r}{N}$ = ... → C = kg/min
PUMP PRESSURE WHEN MUD W_2 REACHES THE BIT: $w_2 = w_1 + \Delta w$ = ; $PR_2 = PR_1 - \Delta PR \times \Delta w$ = kg/cm2
PUMP PRESSURE WHEN MUD W_3 REACHES THE BIT: $w_3 = w_2 + \Delta w$ = ; $PR_3 = PR_2 - \Delta PR \times \Delta w$ = kg/cm2

IV_ Calculation of pump pressure at the end of the control period

$PR_{end} = PR_r + \Delta pc_1 \times \frac{W_r}{W_1}$ = ... + (THROUGH ONE LINE / THROUGH TWO LINES) x ... → PR_{end} = kg/cm2; → PR_{end} = kg/cm2

V_ Check that open hole is not fractured with the chosen 'S' margin

$W_{eqv} = W_r + 10\frac{\Delta pc1}{Z_{weak\ point}} \times \frac{W_r}{W_1}$ < W_{frac}; W_{frac} = kg/ℓ
THROUGH ONE LINE W_{eqv} = + 10 x =
THROUGH TWO LINES W_{eqv} = + 10 x =

VI_ Weighting-up the mud in the riser (before opening the BOP's)

RISER VOLUME V_r = ℓ
REQUIRED WEIGHT OF BARITE $B_r = V_r \times 4.2 \frac{W_r - W_1}{4.2 - W_r}$ = x 4.2 $\frac{-}{4.2 -}$ → B_r = kg

Fig.A.12.1.

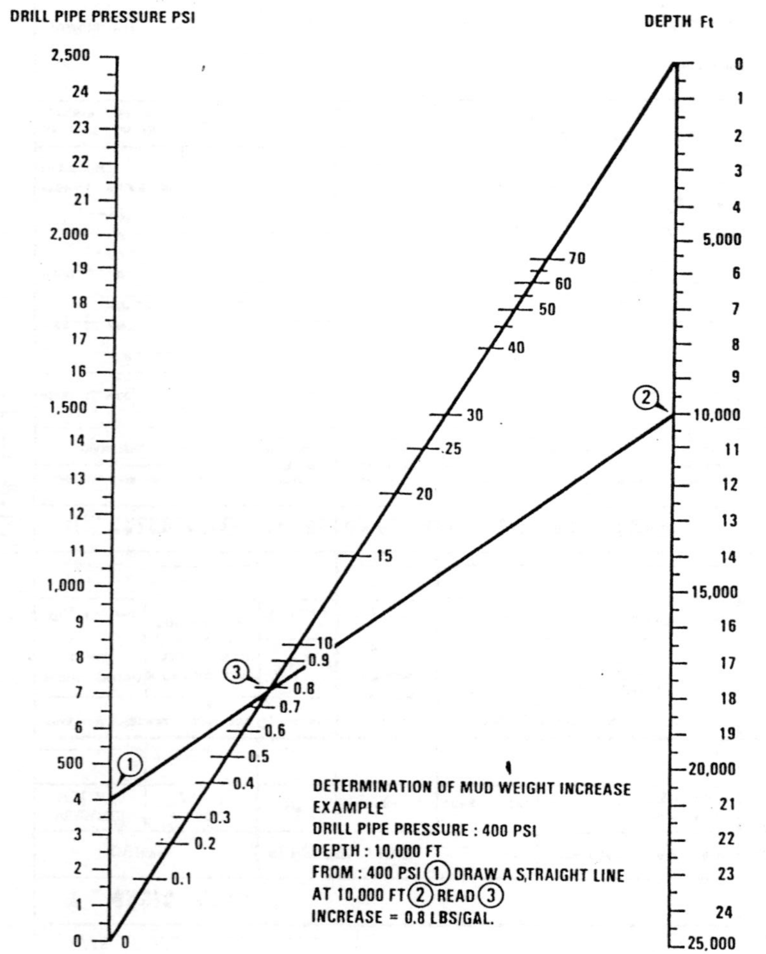

Fig.A.12.2. Determination of mud weight

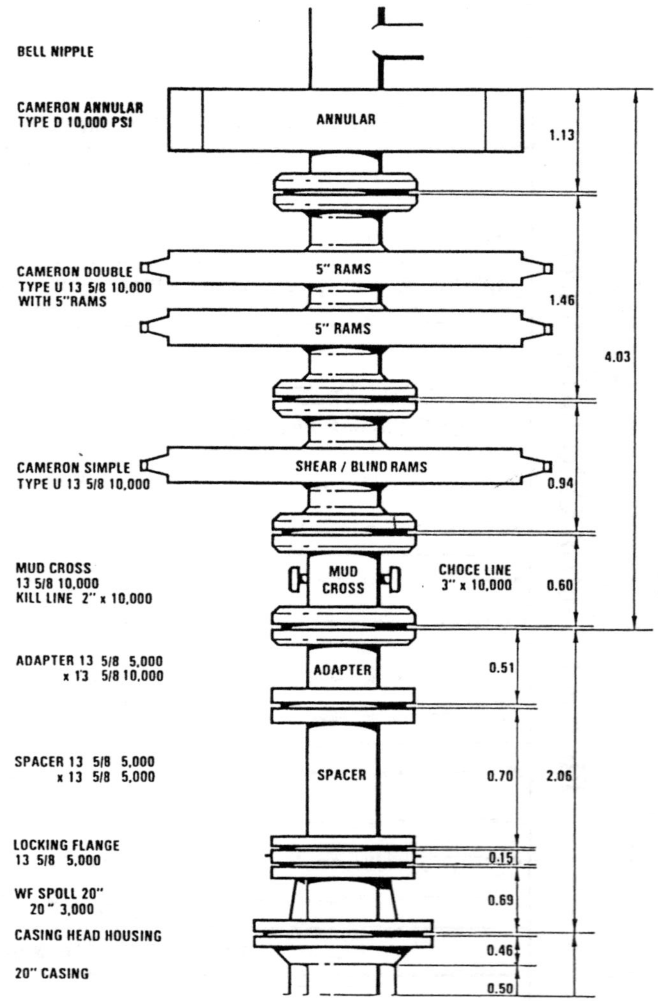

Fig.A.12.3.

Appendix 13

Well Bottom Measuring Equipment

(This appendix is an extract from Pétrole et Technique, No.285, December 1981.)

It is now possible to measure certain parameters at the bottom of wells as a result of recent progress in technology, particularly in the fields of electronics and precision mechanical engineering.

The purpose of this appendix is to review the well bottom measuring equipment recently developed and brought onto the market following the formidable increase in the number of development drilling projects over the past ten years. Two major categories of equipment, now classic, have been introduced since 1970, and are now widely used:

(1) systems with a guide cable inside the pipes;
(2) systems with mud flow pressure modulation.

We shall be dealing only with the most widely used devices well past the laboratory stage and consequently accepted by the drilling industry on grounds of their proven resistance to the hostile environment at the well bottom.

CABLE TYPE ELECTRICAL TELEMETRY SYSTEMS

This system of telemetry is extremely attractive in view of the following advantages:

- no energy source required at the well bottom;
- instantly suitable for two way communication;
- it can transmit even in the absence of circulation, e.g., when a closed well is under observation;
- it has the highest transmission capacity.

Rotary Drilling Cable

The major disadvantage here is the interference on the part of the electrical conductor(s), with the discontinuous process of rotary drilling, where pipes are added singly as the well is deepened. Most attempts to overcome this disadvantage using special pipes comprising an insulator conductor section have not been commercially successful in view of the difficulty in maintaining a quality electrical connection and good insulation at the drill pipe joints.

The Karcher and Ameraga systems were tried out some 40 years ago. More recently, Shell systems (now under licence to N.L.) and Godbey systems have resumed this principle and have enabled measurements of an extremely high quality to be obtained: unfortunately, specialisation of the drilling shaft does not allow widespread use, given the reticence on the part of drilling contractors who in most cases own tubular equipment. The other method is that of a continuous cable, passing inside the pipes, and this is difficult to render compatible with rotary drilling operations. The Exxon system uses a reeving system inside the rotating drill pipes to accommodate the length variations on each addition. The top of the cable can be disconnected for addition of drill pipes, and connection with the outside is provided by turning contacts at the top of the kelly.

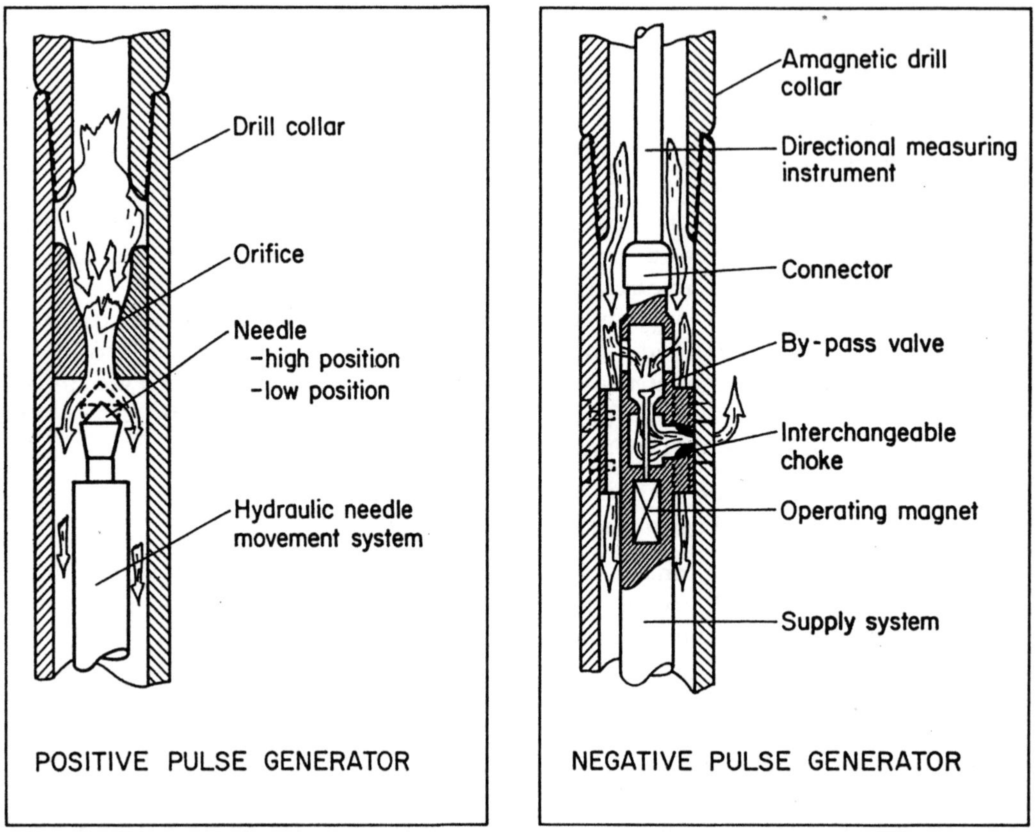

Fig.A.13.1. (a) Positive pulse generator, (b) negative pulse generator.

Table A.13.1.

Transmission processes	**Transmission media**	**Well bottom environment**
1. Sound (or mechanical) waves 2. Electromagnetic waves 3. Variation in hydraulic resistance of the mud system 4. Electrical signals	1. Steel in string of drill pipe 2. Ground around the well 3. Mud inside the string of drill pipe 4. Cable or conductor included in the string of drill pipe	. Very high ambient pressure (300 bars at 2000 m with d = 1.5) . High temperature (normal gradient + 30°C per 1000 m) . Relatively unknown vibration and violent impact conditions (acceleration over 10g) . Abrasive and corrosive fluid (presence of dissolved gases) . Reduced dimensions with extremely limited diameters

The measurements obtained by Exxon are of extremely high quality, but the process is not yet ready for marketing as it is not possible to carry out operations inside the drill pipes (breaking down) due to encumbrance by the complex cable holding equipment.

Cable in Drilling Without Drill String Rotation

The field of application for cable telemetry systems is therefore essentially in drilling with a well bottom motor, where the drill string remains stationary.

The measuring probe is positioned in a special fitting (magnetic drill collar or sensor mounting) and connection is via a continuous cable issuing from the top of the kelly. The measuring probe is brought up with the cable, or may be permanently installed at the well bottom in the sensor carrier fitting (Silvermann system tried in 1940). In this latter case, preferred for measurements in the annular space or on the ground, a connector which can be inserted into the mud on each cable trip is required.

Numerous steering tools successfully use cable telemetry in the build-up phases of directional drilling operations to control both the azimuth inclination parameters and the orientation of the bent sub above the turbine.

The weak point used to be the cable, which had to undergo a return trip whenever drill pipes were added. A major improvement has recently been made with the side entry sub, allowing pipes to be added, from a certain level, without being hampered by the cable, and thus enabling drilling to be carried out with the aid of the bottom motor throughout the life of a tool without need to touch the cable.

The Institut Français du Pétrole has developed and marketed a magnetic steering tool known as the Azintac, which we shall examine in detail. This device, which is extremely useful for build-up, uses an entirely static measuring sensor with three acceleration meters and three magnetometers positioned by a mule shoe in an amagnetic drill collar. The signals are processed at the surface and the instrument has the advantage of allowing direct calculation of the necessary rotations on the surface for changing the azimuth and inclination.

MUD PIPE TELEMETRY SYSTEMS

The first inventors such as Irwin (1943) and Arps (1950) thought that vibration, even of a high frequency, could be easily transmitted through the moving column of mud, at the speed of sound over the greatest lengths. In reality, absorption is high in view of the materials suspended in the mud, and, in addition, the presence of the drill pipe joints periodically modifies the flow conditions, causing numerous stray vibrations. It was soon found necessary to cause a restriction of the column of mud to obtain a signal perceivable at the surface. The first practical invention is attributable to the Varney Brothers (1943, 1946, 1961), and led to the introduction to the market of a very simplified drift indicator known as the Teledrift. This is still used today in Canada and Austria, with a certain degree of success.

Based on the ideas of Arps (1956) and Alder (1959) an attempt was made to market a drill logging system at the beginning of the 1960s. But the true market having provided the impetus for development efforts was directional measurements on offshore platforms.

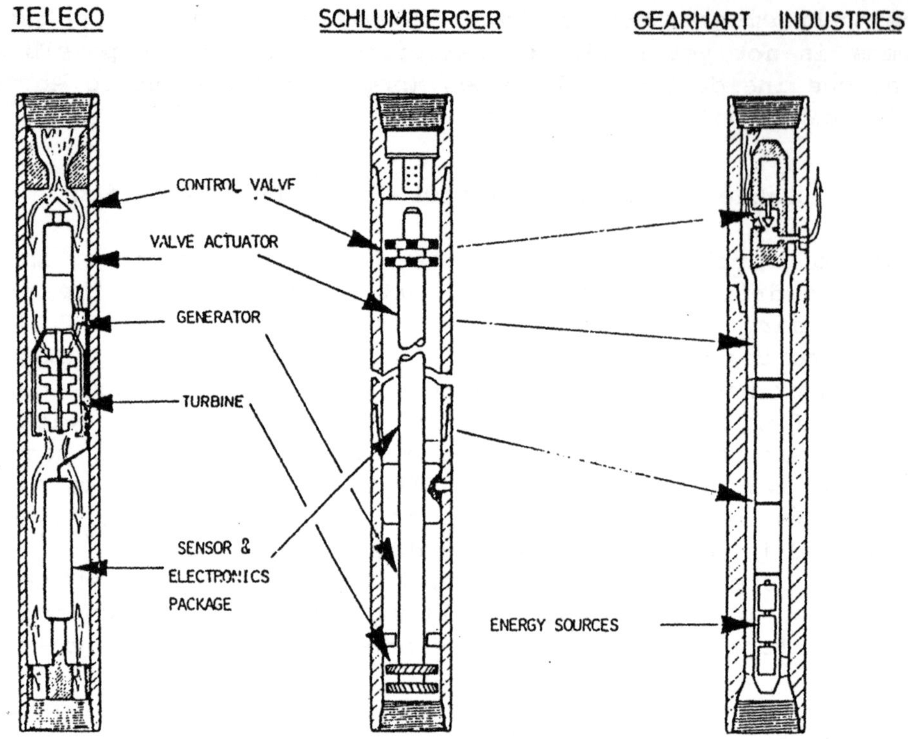

Fig.A.13.2.

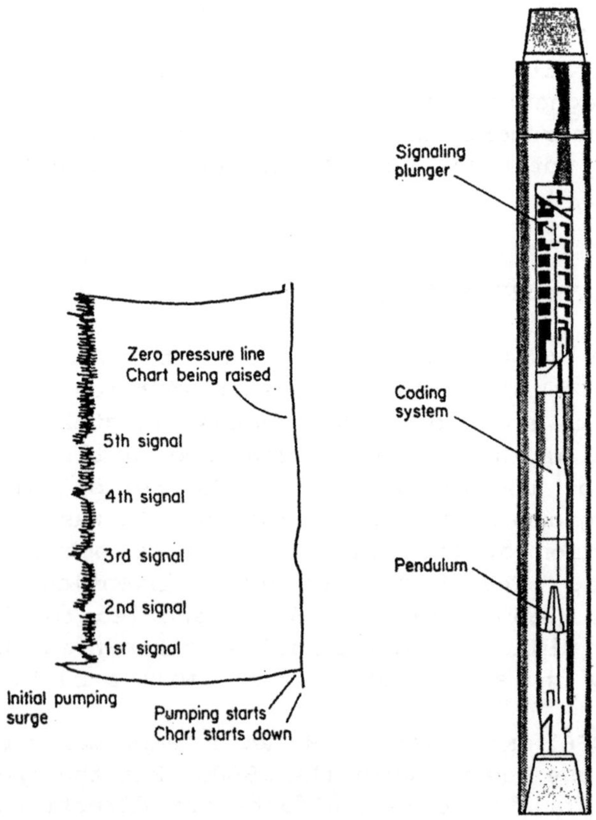

Fig.A.13.3.

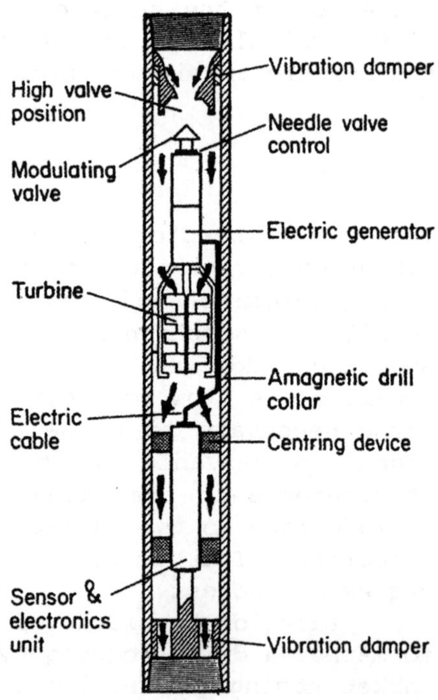

Fig.A.13.4.

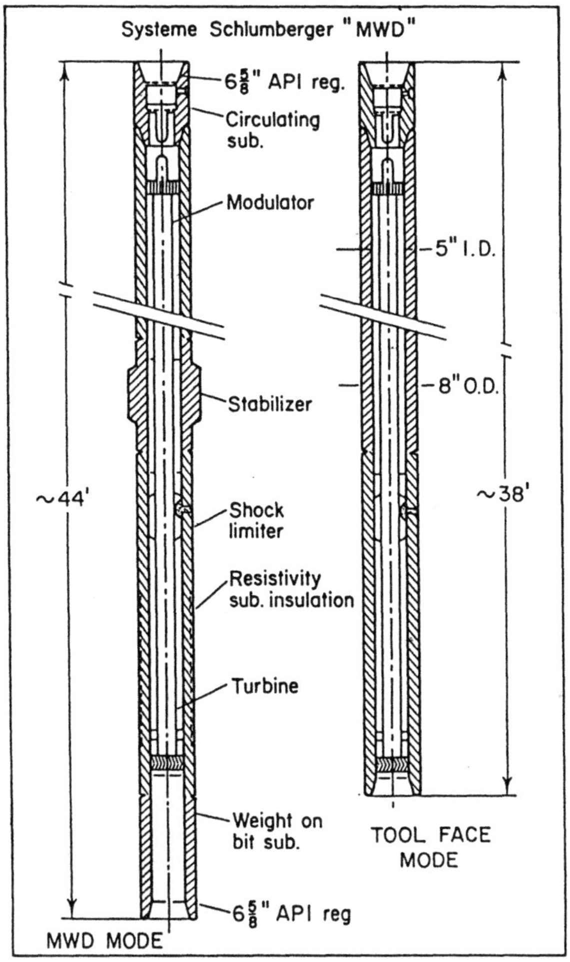

Fig.A.13.5.

At least three companies are active in this field, with slightly different technologies: these are Gearhart Industries, Schlumberger - The Analysts, and Teleco.

The essential advantage that they have over all other systems is the elimination of the need for an additional special purpose link between the well bottom and the surface, at the same time keeping the internal route of the drill pipes free, i.e. they are totally compatible with rotary drilling. The telemetry system essentially comprises two parts:

(1) a bottom transmitter including measuring sensors and comprising a special purpose drill collar located in the vicinity of the tool;

(2) a surface detector comprising one or more pressure sensors connected to the mud injection pipe and an electronic decoding and display system.

We shall now review the various necessary items in such systems, highlighting their essential features.

Transmitter

The principal function of the transmitter is to produce signals in the mud circulating inside the drill pipes. This is done by modulating the size of port offered to the fluid in a special valve crossed by the entire flow of mud. This valve is located at the top end of the drill collar containing the transmitter.

A secondary function of the transmitter is to supply energy for the well bottom instrumentation. Batteries or cells were used at the beginning of experiments, but were found unable to withstand temperatures exceeding 80-100 degrees centigrade, and they were relatively sensitive to vibration and impact. Thus the power required at the well bottom is taken from the column of mud by means of a small turbine together with an alternator. Lithium batteries have, however, been improved for high temperature duty, and constitute a solution for the supply of electrical power to low consumption well bottom systems (intermittent measurement). In certain cases, a hydraulic pump supplies the pressurised fluid necessary for the modulation valve.

The basic components of the transmitter are the modulating valve, the valve actuator, the alternator, the compensation system, the joint or packing, the turbine, the connections to the sensors, and the suspension systems.

The Modulating Valve

This item varies between the different competing systems. The needle valve system is the oldest and was used by Varney (Teledrift), Arps (Logging Tool) before being adopted by Aquitaine and Teleco.

The needle is situated in the centre line of the transmitter opposite an orifice through which the entire mud flow passes. In the retracted position, there is practically no pressure drop. In the advanced position limited by a mechanical thrust block, the restriction on the flow of mud creates a pressure drop of several bars, perceivable at the surface. The system is simple and reliable, but requires a high level of power. There is no risk of plugging, and in the event of instrumentation breakdown, the mud pushes the valve to the resting position.

The production of a continuous wave, although possible with the aid of a controlled needle valve, is easier with a 'siren' comprising an assembly of two wheels with the same number of slots, one fixed and the other rotating, crossed by all or part of the mud flow. This is the system adopted by Mobil/Schlumberger: it has the advantage of a reduced power requirement for the valve, hence simplification of the instrumentation as a result of using an electric drive.

A third system for producing pulses consists of periodically discharging a limited fraction of the mud flow rate causing a negative pulse at the surface. The power required to actuate the valve remains limited, but the reduction in pressure drop actually obtained is dependent on the downstream circulation conditions, such as the presence of a bottom motor or jet tool. Gearhart Industries have perfected this type of valve to make it highly reliable and a low consumer of energy (use of batteries).

Valve Actuator

Depending on the power required, the actuator may be a simple magnet, an electric motor or a more or less sophisticated hydraulic system providing a high power under a low volume.

Alternator

The alternator supplies the electrical energy necessary for all bottom instruments. The first systems were not regulated and had to cope with broad variations in speed of rotation to supply powers of several hundred watts, with extremely severe dimensional constraints for the diameter. This difficulty is currently reduced by the introduction of modern electronics with a low consumption of energy.

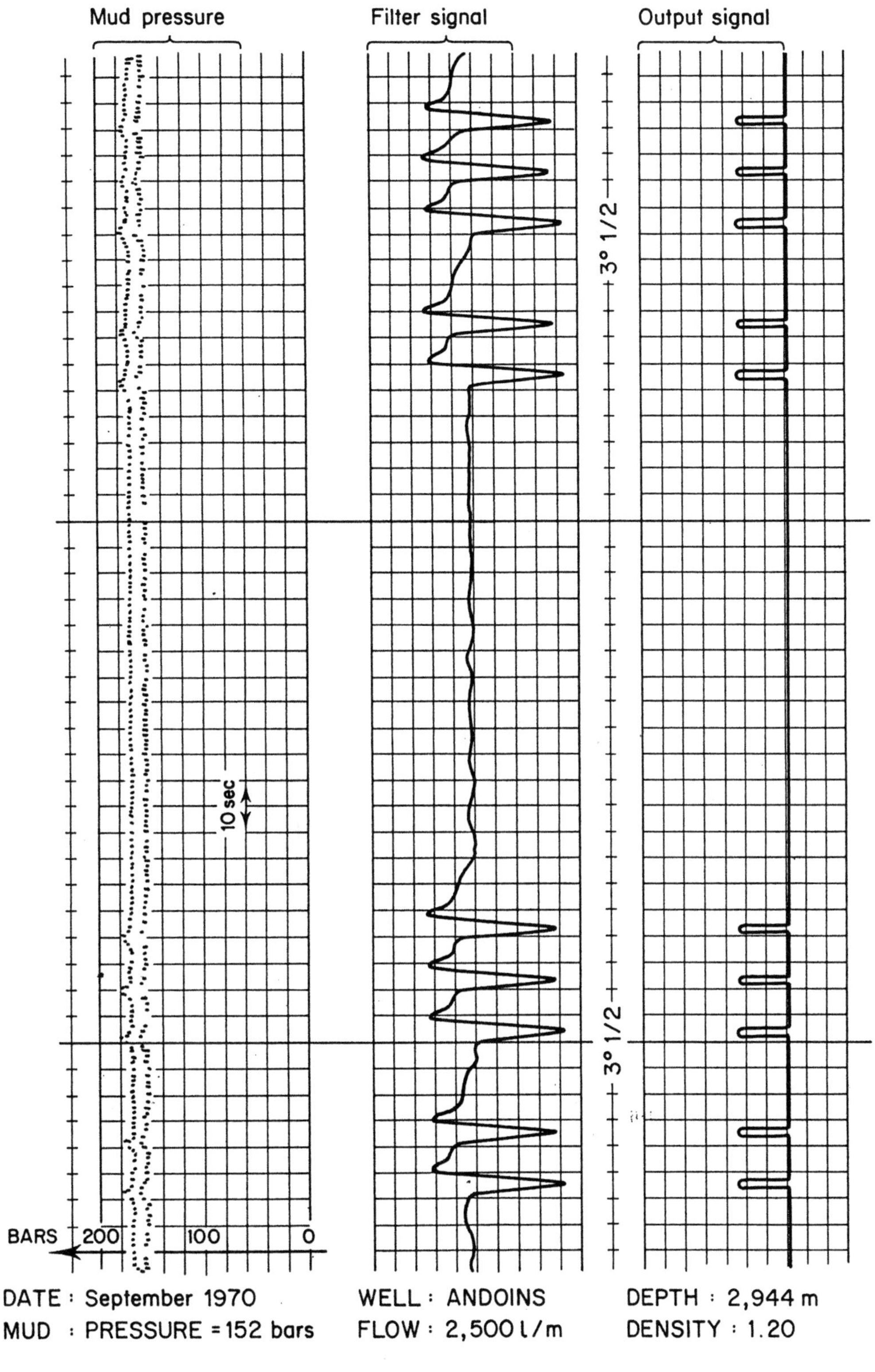

Fig.A.13.6. Inclination measurement while drilling

Compensator

The valve actuator and alternator are normally bathed in a lubricating fluid at the ambient pressure of the well bottom, to avoid the enormous differential stresses on the shafts and axles to the turbine and the valve which have to be placed in the mud.

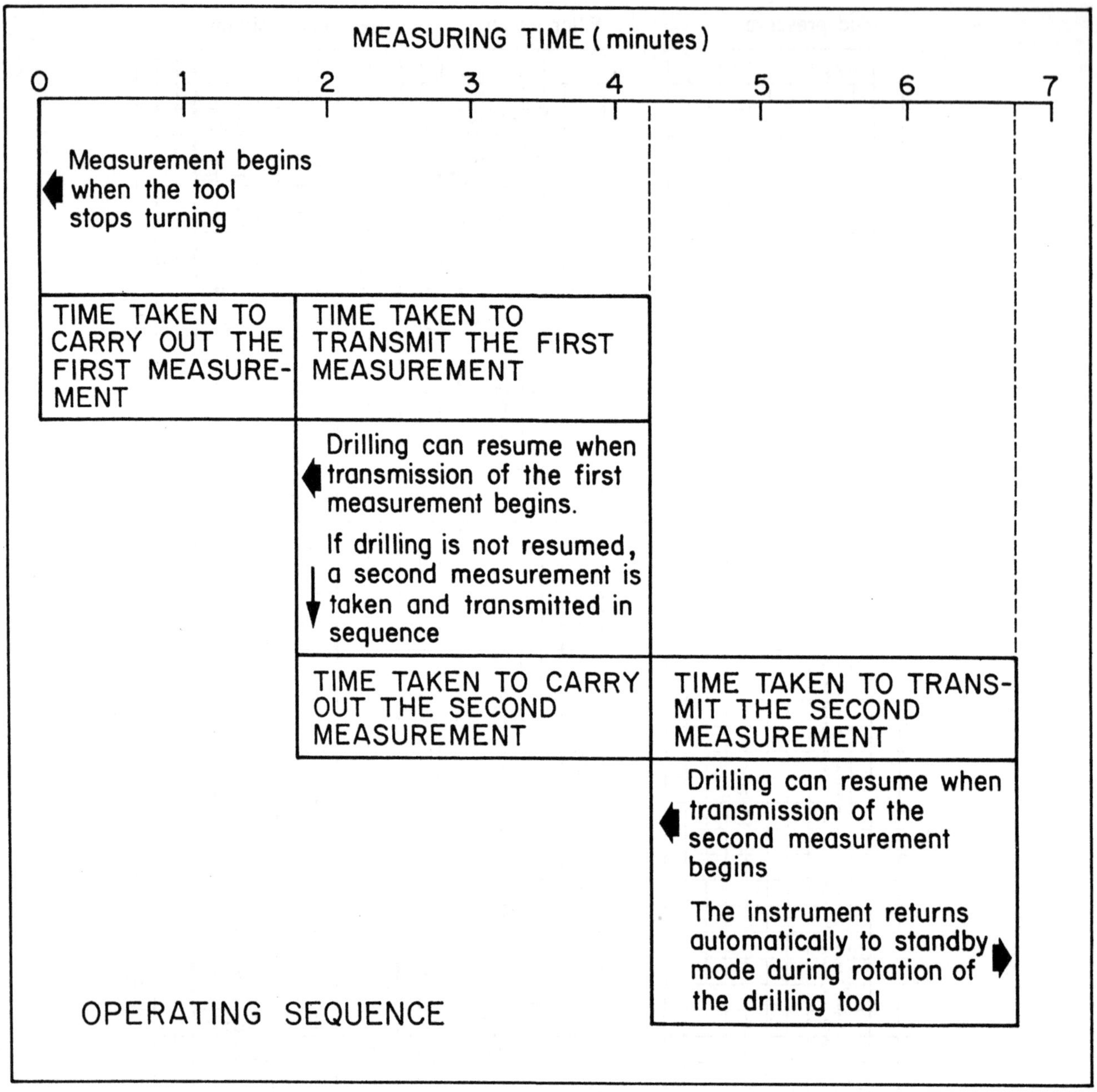

Fig.A.13.7.

The role of the compensator is to equalise the pressure while allowing the lubricating fluid to expand or be compressed under the effect of the temperature and pressure. This is achieved by diaphragms or bellows, which have to remain flexible and leak-tight while at the same time withstanding the thermal and chemical effects of the mud (additives, acids, gas, etc.).

Rotating Joint

This joint separates the fluid lubricating the well bottom instrumentation from the abrasive drilling mud. In most cases it is a flat joint in abrasion resistant material; this technique is also being used in cement works and china factories.

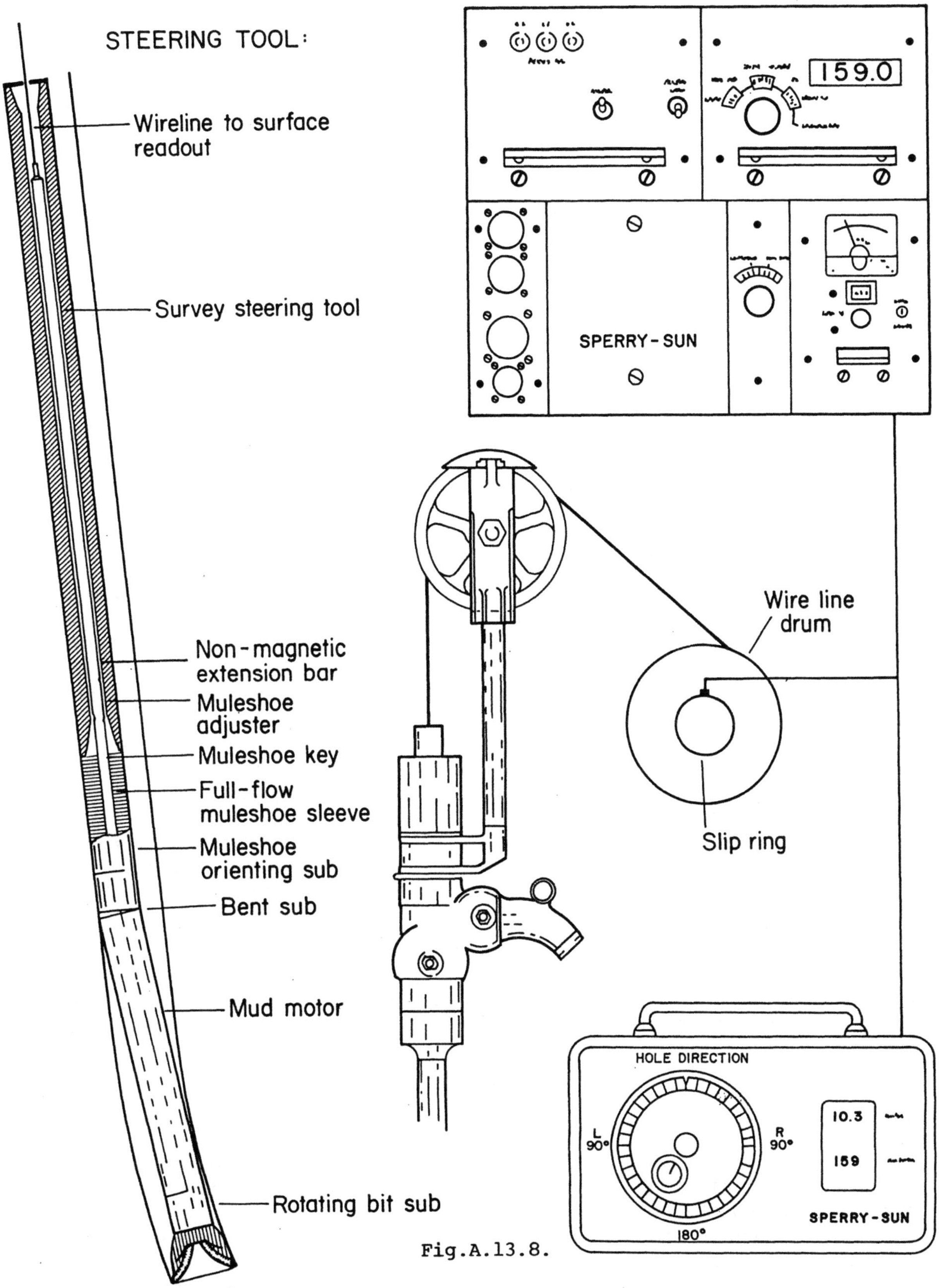

Fig.A.13.8.

Turbine

The turbine has around one to three stages, similar to those of the drilling turbines. The mud flow rates must be capable of varying between 1000 l/min and over 3000 l/min, to cover the fields in which measurements are required during drilling. As turbine power varies as the cube of the flow rate, a number of types must be provided. The turbine is generally suspended from the lower part of the well bottom equipment, with the ball thrust and bearing in the oil.

Connection to Sensors

Sensors fall into two categories, depending on the level of connection required:

(1) those measuring a physical quantity present inside the tube containing the equipment;
(2) those measuring a physical quantity requiring contact with the drill collar or the environment.

The former do not require any connection in the mud and may be included in the processing and coding electronics (e.g., inside temperature, gamma ray and inclination).

The other sensors are located on the drill collar containing the equipment and require electric wires to be run to the atmospheric pressure enclosure containing the signal processing and coding electronics. Dismantling constraints make these connections in the mud fragile and somewhat unreliable in the drilling environment.

Hanger Device

The electrical components and mechanical parts of the sensors and transmitters cannot be permanently exposed to the impact and vibrations caused by drilling. It is therefore recommended for all instrumentation to be suspended inside the drill collar by means of dampers, generally of the elastomer type, capable of retaining satisfactory properties under high ambient pressures and temperatures.

The use of these dampers must not disturb the operation of the transmitter valve and must be compatible with the connections to sensors located on the drill collar (e.g., weight, torque, mud and annular temperature and resistivity).

Particular attention must be paid to this point in the design of a well bottom transmitter, since it has a considerable influence on the service life and commercial success of the entire system.

Surface Instrumentation

The surface installation must be suitable for insertion on any drilling rig without hampering the personnel responsible for directing operations. It comprises, as a minimum, one or more pressure sensors and an electronic decoding system with a read-out of the quantities measured.

Pressure Sensors

The variations in hydraulic resistance caused at the bottom of the well by the transmitter are translated into pressure variations at the surface, and may be sensed on the swivel hose supply line.

For pressure pulses repeated every 3 to 5 seconds, a single sensor located on the fixed piping will suffice. In the case of a continuous wave with a higher frequency (12 Hz for the Schlumberger system) it is necessary to eliminate the stray noise caused by the drilling pumps in the same frequency range and two pressure sensors are sometimes necessary to provide a reliable reconstitution of the signal from the transmitter.

In all cases, it is recommended that the gas pressure of the pulsation dampers be adjusted to a value equal to 60% of the injection pressure, with a maximum of 70 bars, to regularise the drilling pump rate.

Surface Decoding

Decoding may be autonomous (Teleco) or may form part of the surface measurement system (mud logging). In the former instance, the decoding system is supplied in the form of one or two modules, at least one of which may be placed near to the drilling foreman's position, to display the values measured at the bottom (directional drilling). These modules are protected and brought in line with the safety standards by continuous purging with inert gas.

When well bottom measurement services are provided by the same company as surface measurement, a cabin is provided to house all display and calculation instruments and the pressure sensors should be connected to this unit, providing for signal decoding and read-out, possibly by computer.

APPLICATIONS IN DRILLING OPERATIONS

The potential applications for the transmission of measurements from the well bottom to the surface are numerous, for both development drilling and exploratory drilling. The object of this article is not to present the various sensors which may be used at the bottom of the well for in situ measurements. Everything measurable by logging cable could conceivably be evaluated while drilling, and the limitation arises only in the physical possibilities of adapting and fitting sensors of the same type on the lower part of the drill string. In addition, the transmission system offers the possibility of measurements on the drilling operation per se. For example:

- the speed of rotation of the well bottom motor;
- the weight and torque on the tool;
- vibration and acceleration of the string of drills;
- orientation of the motor used with a bent sub.

Measuring Frequency

In very approximate terms, the need to obtain measurements may be regarded as proportional to the rate of advance. The spacing of the 'SN' resistivity measuring electrodes is 16 in., or 40 cm, generally sufficient for evaluating the formations crossed. If the rate of advance is of the order of 12 m per hour (as in the case of soft ground), each parameter must be measured every two minutes. On the other hand, when the advance is only one metre per hour (in extremely hard ground, using a worn tool, etc.), only one measurement is required every 24 mins. to obtain results comparable with conventional logs.

For directional measurements, three parameters are frequently necessary (inclination, azimuth and tool facing or reference), but the measuring frequency is not more than one point every metre in a kick-off phase with a bent sub and turbine, falling to one point every 10 m or on every stab in the stabilised inclined phases. Taking into account the three measurements required, the three parameters have to be transmitted every six minutes or every hour respectively for a rate of advance of 10 m per hour. In reality, in the stabilised rotary phase, attempts will be made to limit the stoppage time for rotation of the drill string at the end of the drill pipe to carry out measurement prior to stabbing.

Dynamic measurements including, for example, torque or acceleration at the tool, require a greater number of sample measurements to be transmitted. If average values are required over 30 sec., it may be regarded as sufficient to transmit six measurements per minute (three parameters). On the other hand, for instantaneous values covering only fundamental modes of vibration of the drill strings, excluding any harmonics due to the action of the head wheel teeth, measurements must be taken 30 times per second for each parameter, giving 5400 to be transmitted per minute.

Hence the means of transmission from the bottom to the surface must be suitable for the measurements which have to be taken at tool level, cable being preferable for the transmission of dynamic measurements. However, the pressure modulation system will generally be suitable for the requirements of drilling and for evaluating formations, taking into account the average rates of advance from a certain depth. Progress in the bottom processing of dynamic measurements is desirable and will probably be made in the near future with the aid of miniaturised data processing systems.

I leave it to the representatives of the companies attending this conference to describe the specific sensors used in their rigs, and the procedure for use on the drilling site. They will also provide you with information on the results obtained in wells, with particular reference to service life and anticipated secondary benefits.

Appendix 14

Procedure in the Event of Risk of H2S Leak Whilst Drilling

(Key: for diagram see Fig.A.14.1.)

1 Normal drilling.

2 Circulation loss.

3 Show.

4 Appearance of H_2S in mud (sulphurous) or in the gas extracted.

5 Traces of H_2S in the atmosphere.

6
- Treat the mud.
- Install detectors.
- Reinforce monitoring at the bottom plug and at drill collar pull-out.

7
- Stop drilling and circulation.
- Alert (the mud house
(the site manager
(the inspector.
- Observe the return line from a distance.

8
- Ventilate the floor.
- Check breathing apparatus.
- Connect the vertical separator to the wind side horizontal flare.

9
- Resume circulation, checking the H_2S content.
- Resume drilling after ¼ hour's observation whether H_2S 10 ppm.

10 H_2S alarm.

11
- Shut off the well (BOP and chokes) (3).
- Observe pressures for 10 minutes.
- Circulate via fully open choke and vertical separator connected to horizontal flare treating the mud.

12 Well stable.

13 Losing returns.

14 Show.

15
- After one cycle, and if the H_2S level becomes $<$ 10 ppm, open the BOP and continue circulating, treating the mud.
- Resume drilling when the H_2S has completely disappeared from the atmosphere.

16
(1) The alarm threshold is 10 or 20 ppm in the atmosphere, according to the local regulations.
(2) If a trip loss or show occurs, follow procedures F8 or F9, paragraph 2.
(3) On a floating device, place the drill pipes on BOP immediately on closing in accordance with the instructions F9, paragraph 2.

(4) H_2S = traces of H_2S detected in the atmosphere by fixed or portable detectors, irrespective of contents.

(5) Crater risk is the possibility of the gas flowing behind the casing string in the cement or ground and erupting out of control on the surface. They may occur with shallow or badly cemented casing strings or those covering unconsolidated grounds. It is the responsibility of the drilling department manager, after setting the casing string, to state whether or not there is a crater risk.

(6) On a floating unit, it is preferable to close the well, disconnect the riser and remove the unit.

17 H_2S alarm triggered by the drilling inspector and/or site manager who must inform the rig manager and supervise the following operations.

(a) Evacuation of the mud pit and vibrator room.
(b) Each man in the emergency team to put on a breathing mask.
(c) Other personnel to evacuate to the safety area, with escape masks.
(d) Floor to be given maximum ventilation, opening doors and removing panels.
(e) Tour of floor, vibrators, cellar, pump room, mud house.
(f) Air extractors to be set to full, and doors of pit room closed.
(g) Check on system under chokes and connection of vertical separator gas outlet to the leeward side horizontal flare of the well.

18 Partial loss.

19 Total loss.

20 No leaning over well or ditch without mask.

21 - Stop drilling.
- Halve circulation.

22 - Stop drilling and circulation.
- Close annular type preventer (3).
- Fill well with water using K.L., measure volume.

23 - Alert the mud house, site manager and inspector.
- Check H_2S with a portable detector.
- Ask geologist for H_2S indices.

24 Plugging by customary methods.

25 H_2S alarm.

26 - Stop circulation.
- Close annular type preventer (3).

27 - Circulate with plugging agents, at reduced flow rate, through the fully open choke and the vertical separator connected to the horizontal flare.
- If no return, fill the annular space with water via the K.L.

28 Lack of air.

29 No risk of crater (5).

30 Risk of crater.

31 Alert the mud house, site manager and inspector.

32 - Show control using customary procedures F8-F9.

- Reinforcement of H_2S detection.

33 P_a reaches P_{adm}.

34 H_2S alarm.

35 - Purge the well to keep $P_a = P_{adm}$.
- Circulate weighted mud at maximum rate.

36 Fracturation.

37 Partial evacuation.

38 Light the flare on arrival of the gas.

39 Internal eruption control by squeezing plugging mud in annular space.

40 - Purge and circulate with plugging agents.
- Observe signs of crater.

41 Lack of air.

42 First sign of crater.

43 - Stop pumping.
- Close the well.
- Evacuate contaminated zones.

44 - Open the well on the flare.
- Ignite.

45 Total evacuation.

46 Offshore. The alarm is given by the rig manager responsible for safety in the quarters and organising part or total evacuation in accordance with instructions F11.
Onshore. The drilling inspector:
- advises base and requests assistance from the safety department;
- if the situation so requires, advises the local police or authorities to request them to arrange protection of the local area.

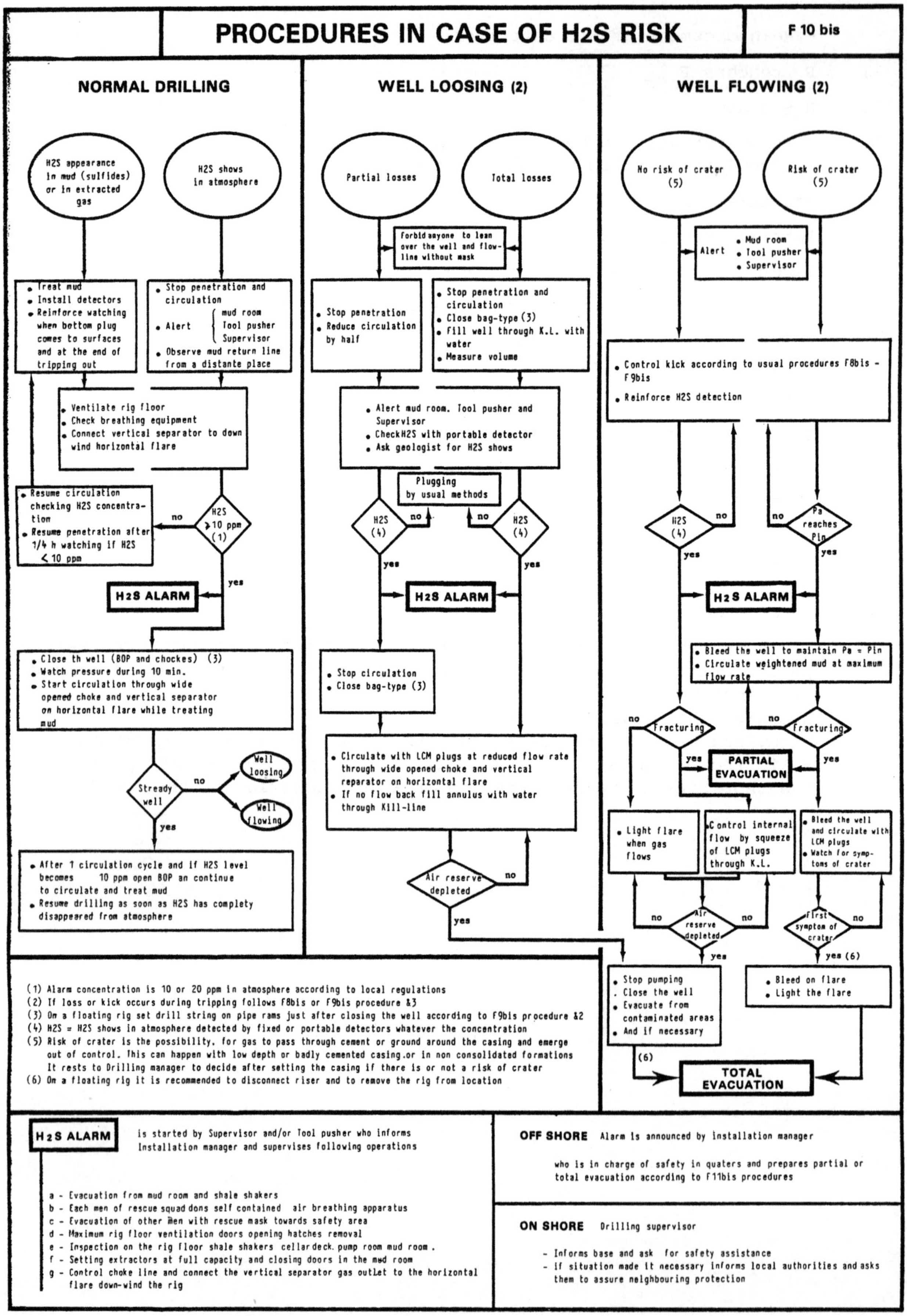

Fig.A.14.1. Procedures in case of H_2S risk.

Appendix 15

Blow-out Control Procedure

PRELIMINARY PHASE

Personnel must be ready to act at all times and, to such end, a number of procedures must be produced and regularly updated.

(1) Emergency team organisation. An organisational chart must be produced and passed to all persons appointed as members of the team (and their deputies).
(2) A schedule of resources available in the area, and where appropriate, outside the area. These resources (high pressure/delivery pumps, BOPs, etc.) are in most cases in the hands of the contractors - hence priority cooperation must be established with these contractors to cover the event of blow-out.
(3) A reaction time scale to cover the period following blow-out alert: delivery times etc. with a view to preparing a PERT type programme of control operations.
(4) Maximum understanding of the well with respect to:
    (a) reservoir plus geology
    (b) trajectory of existing wells and wells currently being drilled
    (c) drilling parameters.

CONTROL PHASE

Control methods may be divided into two main categories:

(1) surface control or 'capping';
(2) sub-surface or relief well control.

Capping

This is possible where access can be gained to the head of the well under blow-out. The outline of the operation is as follows:

- clearing the wellhead (after analysing the existing conditions to ascertain whether it is desirable to close the well at the surface);
- putting a closing assembly in place: BOP + by-pass lines + injection line;
- closing;
- pumping an effluent of a sufficient density to the bottom of the well to control the blow-out;
- generally plugging with cement.

Note. There have been many inventions in the field of capping. Particularly worthy of mention are the methods currently under investigation for offshore use, whereby plugging agents are injected into the well casing (after drilling under pressure and installing a valve), these agents being designed to plug at a higher level where the outside of the casing has been crimped.

Examples: The 'Crimp and Plug' method developed by Tenneco and Brown Oil Tools in the USA, its principle being adopted in the 'Mobell' method designed by the Norwegians.

Relief Wells

Where it is either not possible or undesirable to work directly on the wellhead (as in almost all cases offshore), the relief well method of control is used. This involves moving away from the blow-out well and intercepting it, by drilling a well (or possibly two wells) to the erupting elevation. It does not intersect the blow-out well, being located at a distance of 5 to 15 m. It simply means using the porosity and permeability of the eruptive ground between the two wells to pump from the relief well to the erupting well.

This method may be outlined as follows:

- locate and then drill the relief well(s);
- locate the erupting well, make any corrections in trajectory to reach the target (15 m from the erupting well);
- carry out a communication test by pumping water. Acidify if required (calcareous rock);
- dynamically control the well by injecting water at a high flow rate and pressure (this invades the eruptive rock zone in the immediate vicinity of the well);
- statically control the well by injecting mud (or cement) of a sufficient density;
- place a sealing assembly on the controlled well.

Note. Pumping control or 'dynamic killing' must be based on a simulation study on the migration of water (mud) from one well to the other, as a function of:

- the reservoir rock (porosity, permeability, thickness, etc.);
- the distance between the wells;
- the flow rate and injection pressure at the head of the relief well.

Simulation gives the necessary time for the production of the eruptive well to be at 100% of the water (injected) plus the additional time for the mud to reach the well.

Appendix 16

The Fire Injection Spool System of Blow-out Control

The most dangerous of all accidents liable to occur in oil operations is, without question, a blow-out. Fires as a result of these blow-outs can destroy rigs and cause the loss of numerous human lives (in particular on offshore platforms).

The risk of blow-outs is always present in oil operations (with greater or lesser probability depending on the operations being performed). Numerous items of equipment have been designed and installed on drilling rigs to counter this risk, but sometimes these items (in spite of their increasing in-built redundancy) are defective, and any work to control the well proves impossible either because the effluent has ignited or because personnel would be taking too great a risk by approaching a wellhead with an erupting hydrocarbon liable to ignite at any moment. As an example, we describe below a fire-fighting device that represents a highly realistic approach to the problem.

Following the blow-out of Gassi Touil in the 1960s, safety officers tried to find a system capable of protecting against these risks. Following numerous tests on diverse and varied techniques, they agreed that the injection of water directly into the erupting vein of hydrocarbon was the best way of resolving the problem. A system known as the fire injection spool has thus been developed.

This fire injection spool is, of course, a wellhead safety device enabling a fluid (oil or gas) under uncontrolled blow-out to be rendered non-flammable, and any ignited escape to be extinguished within a relatively short period of time (less than two minutes in all full-scale tests performed).

DETAILS

Principle

Risks of blow-out are always present during drilling or completion operations and, in general, throughout all operations under pressure. If fire occurs, there is a great chance of the well being lost, and the costs involved are always considerable.

This wellhead safety device enables the blow-out fluid to be sufficiently wet to be rendered non-flammable, irrespective of the location and form of leak. This result is achieved simply by injecting an extremely fine spray of water inside the wellhead through an injection spool. Numerous tests performed have shown this to be capable of preventing any ignition of the erupting fluid or to achieve rapid extinction (in less than 90 seconds) in the event of outbreak of fire before the device was used.

Since the injection spool is also mounted under the BOPs and the user is therefore guaranteed against any lack of tightness of conventional safety devices, the interest of protection such as this becomes clear: it will contribute to making exploratory drilling less hazardous, and will considerably reduce the risks involved in certain operations, such as work-over or snubbing.

This process will also allow offshore production platforms and the wellheads to underground LNG or LPG storage systems to be protected.

Description

The spool may be fitted to any drilling wellhead. It comprises:

(a) an outer tubular casing with mounting flanges at each end;
(b) a central zone smaller in section at mid height than at the ends;
(c) a circular injection chamber between the central zone and the tubular casing linked to the injection pump by HP pipe and to the central zone by a series of orifices fitted with valves, acting as diffusers when the device is in action, and otherwise as non-return valves.

The system is illustrated in Figs.A.16.1 and 2.

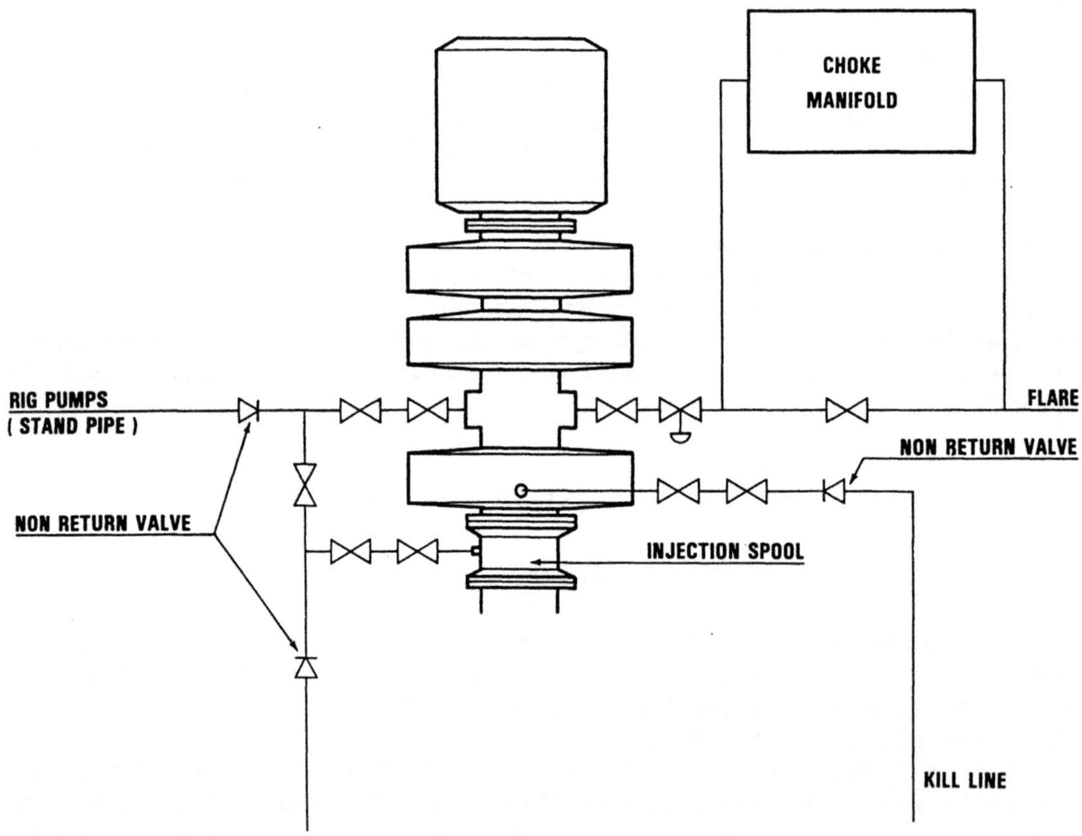

Fig.A.16.1.

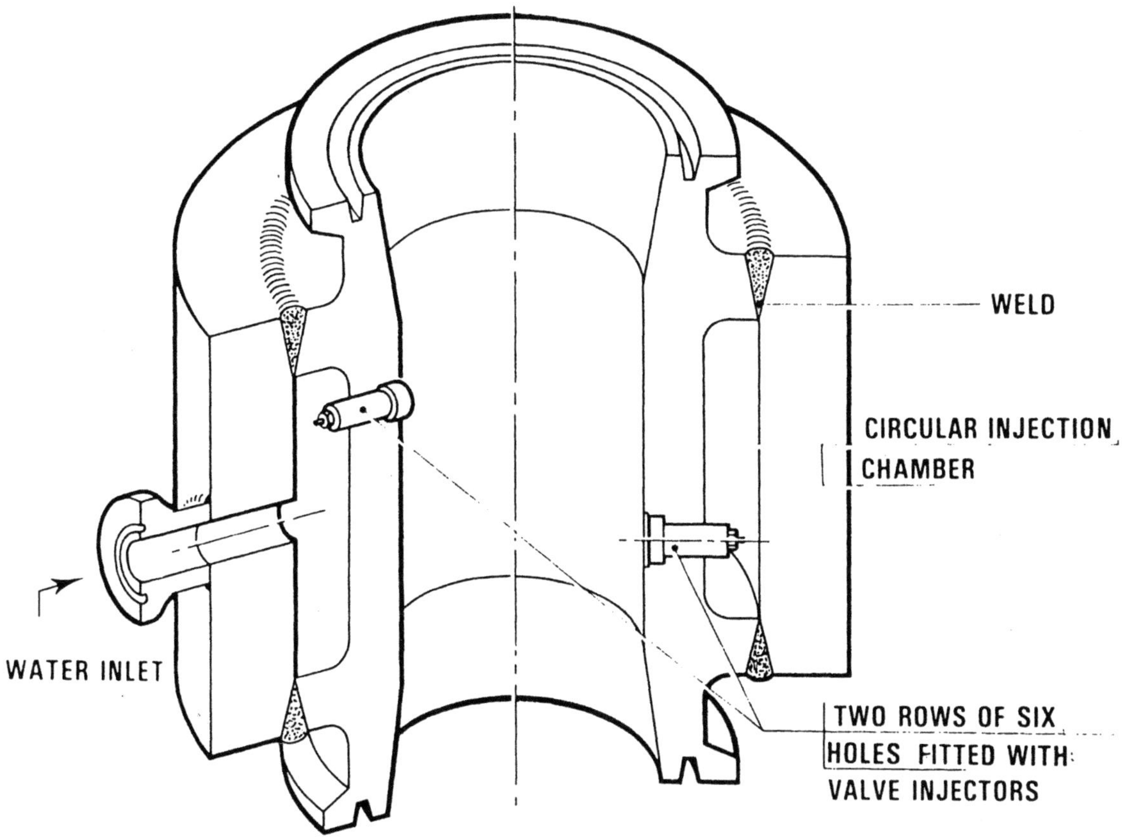

Fig.A.16.2. Prototype 13⅝, 5000 series.

Appendix 17

Living Quarters and the Safety of Personnel

We shall begin by dividing the market into segments, stating the number of platforms in each segment and the specific problems involved in introducing a new evacuation concept. These various segments are: semi-submersibles, jack-ups, and fixed platforms.

SEMI-SUBMERSIBLE PLATFORMS

Number in service: approximately 120, plus around 50 under construction. New construction: same trend as for jack-ups. Semi-subs designed for extremely difficult offshore conditions are likely to increase in numbers.

Their main characteristics are:

(a) major list;
(b) drilling floor at the centre of the platform;
(c) systems must not cause major modifications in the environmental stresses, otherwise the performance of the platform will be affected.

JACK-UP PLATFORMS

There are over 400 jack-up platforms in service, plus around 50 under construction. New projects: very few for the next 2-3 years, but the market distribution is then expected to follow the overall trend which has applied for the past twenty years or so; i.e., increase of approximately 10% in the total number.

New platforms planned are scheduled to work throughout the year in the North Sea and to be used for production.

Problems involved in the introduction of any new concept:

Existing platforms. These are mobile. They could therefore be modified at a yard at reasonable cost. Modifications would only, of course, be feasible on condition that they do not modify the overall design of the platform.
New platforms. New ideas of any kind could be incorporated at the design stage.

FIXED PLATFORMS

There are approximately 4000 fixed platforms in service (with permanent personnel representing several hundred thousand people), plus approximately 200 under construction, including around 30 for the North Sea.

For new projects market trends are difficult to foresee. However, around 50 platforms are envisaged for the North Sea for 1984 and platforms are generally tending to be installed in increasingly difficult seas.

Problem of Introducing a New Concept

Existing platforms. Problems would be extremely difficult to resolve, for two main reasons: (1) the platforms are overloaded, and (2) any work offshore is extremely expensive. New concepts would only be acceptable, therefore, if they involved a minimum of adaptation work on the platform, with no structural modifications.

Platforms currently under design. Here there is no real constraint.

Any new solutions must be approved by the various government authorities (MMD, DOT, USCG, MM) and accepted by the classification bodies and operators' safety departments.

THE CURRENT STATE OF THE ART

Offshore platforms are equipped with evacuation systems derived from ships' lifeboats: life dinghies, or better, Whitaker type capsules. Built in fibre glass, they are launched by cable and operated by means of special keyboards. Their capacity is around 50 persons per capsule.

Platform evacuation may be broken down into four main operations, from the moment at which the platform manager gives the order to evacuate, to such time as personnel are located at a safe distance from the incident:

(1) embarkation;
(2) launch;
(3) distancing from the platform;
(4) waiting for assistance.

The constraints on each operation are summarised in the evacuation table (Table A.17.1).

This shows accessibility of emergency systems and waiting for assistance to be satisfactory.

Unfortunately, however, the same is not true with respect to launching and distancing from the platform. These operations are lengthy, and virtually impossible in bad weather. These weaknesses can only be surmounted by specific improvements. They are in fact directly related to the concept adopted.

Table A.17.1. Platform evacuation by life capsules

Evacuation phase	**Assessment**	**Comments**
Embarkation		
Speed	Good	Embarkation at deck level.
Accessibility	Hazardous or good	Platform specially arranged. Extremely accessible. Out of wave attack range.
Reliability	Function of list	
Launch		Cable systems
System reliability	Mediocre	Cables can become tangled: craft jammed at mid height.
Method of external release	Poor	One person has to sacrifice himself to launch.
Speed of launch	Difficult to control	Too fast: crushed vertebrae. Too slow: pendulum movement.
Risk of damage in launch	Substantial	Capsules knock against the platform.
Distancing from platform		
Separation	Mediocre	Difficult to unhook.
System reliability in storms	Mediocre	Motor can fail to operate.
Manoeuvrability	Very poor	Under storm conditions, the engine is inadequate. In addition, the form of capsule makes it difficult to steer.
Risk of damage	Substantial	Waves can cause the capsule to crash against the platform.
Waiting for assistance		
Stability	Good	For circular capsules.
Reliability	Reasonable	Care with life boats.
Fire resistance	Adequate	Appears to be satisfactory from tests. N.B. Capsules can move away from any water surface fire.

CONCEIVABLE CONCEPTS

Irrespective of the type of platform, the life-saving system comprises three interdependent sub-systems:

(1) access to the craft;
(2) the life-saving craft itself;
(3) the means of launch and distancing from the platform.

These sub-assemblies must comply with the following fundamental criteria.

(1) Ease of access to the life-saving vehicle, irrespective of weather conditions (sea, wind, currents, etc.) and the platform's list.
(2) Positioning and distancing in relation to the platform, to prevent any impact with its superstructures or with any other satellite installation. Distancing from the wreck area must be achieved in the minimum time.
(3) Capacity - recommended as 200% of platform personnel.
(4) The time between the order to evacuate and the time at which the crew are away from the wreck area must not exceed 10 minutes.

These criteria have led us to select three systems as possible concepts: (1) submersibles, (2) evacuation satellites, and (3) launching ramps.

Submersibles

Since the main problems are launching and severe weather conditions, the submersible could prove an effective solution. The submersibles concept is particularly suitable in the case of semi-submersible and fixed platforms. The craft could be installed on the underwater structures of platforms with access via leak-tight ducts run inside the platform, and thus providing for various routes, depending on zones damaged.

The submersible is to be propelled and ballastable. Its propulsion is to be sufficient to remove the craft a few hundred metres from the disaster zone. It will then re-surface.

It is also conceivable for a confined space to be provided inside the platform, providing victims with survival conditions in an hermetically sealed compartment, capable of withstanding an external pressure of 30 to 45 bars (in the event of platform submersion). This procedure would enable victims to wait for assistance in a platform having collapsed. The emergency procedure could be of the same type as that used for the recovery of military submarine crews from the sea-bed (a submersible will clamp onto the platform survival compartment).

Both systems (the escape submersible, or confined space within the platform) require a suitably designed platform. They are relatively unsuitable for adaptation on existing platforms. The possibility of adapting this principle could be investigated for existing platforms and jack-ups.

Evacuation Satellites

The object of evacuation is to transfer personnel to a survival craft, enabling them to await assistance in total safety.

The term 'satellite' is used here to describe a craft which, being independent of the platform, is normally situated a few metres away from the platform, to allow personnel to transfer easily by walkway, toboggan or any other system. This craft must be capable of moving away from the platform automatically and remaining in a survival position at an adequate distance.

This satellite should remain operational under the worst possible weather conditions experienced at the platform. In all cases, the transfer and distancing process must be reliable and easy.

Two concepts could be investigated:

(1) A buoy held by pre-tensioned cables comprising three parts: a post or deck, a riser, floating gear. This is to be kept in 'semi-submersible' position by a cable connected to ballast. Under normal weather conditions, the buoy would be held against the platform by a complementary system, be de-activated in the event of evacuation. The buoy equilibrium position would be the survival position, at an adequate distance from the platform (50 to 100 m).

(2) An articulated column, which would be a long-limbed structure resting on the sea-floor, with a universal joint fixed to a gravity base, or with piled foundations.

The following comments might be made with respect to potential applications for the satellite system.

(1) The system would be tied to the site, and would not therefore be effective in the event of drifting of the platform.

(2) Installation of this system would involve a relatively major operation. Exploratory platforms, changing site on average every six months, may find this too constricting.

(3) This type of evacuation system would involve additional congestion of the oilfield.

In the case of fixed platforms, this heavyweight evacuation system would in fact meet requirements.

Launching Ramps

A launching ramp would allow life craft to be launched by gravity slide (requiring only an initial impulse) or assisted by an autonomous advance system. The ramp could be a metal lattice structure (beam type) ending in a float of a specified volume, with an articulation system (if necessary) to connect it to the platform.

Appendix 18

Safety and Reliability Aspects to be Taken into Account in the Development of TLPs

INTRODUCTION

Reliability/safety analysis on new concepts involves two levels of difficulty, concerning two categories of risk event:

(1) events which are quantifiable by a probability law;
(2) events which are difficult to quantify, in view of the innovative nature of the concept.

The second category of risk event has to be minimised by reduced and/or full scale tests: this constitutes a brake on the use of new concepts, and it is generally in this area that development efforts need to be made.

APPLICATION WITH REFERENCE TO TLPs

(a) Quantifiable events

The elements involved are:

- sea/meteorological conditions;
- modes and processes used in manufacture and in the control of manufacture;
- behaviour of the platform afloat, on tow.

(b) Events which are difficult to quantify

These elements relate essentially to the tension legs, the behaviour of the platform in installation and under operating conditions, and the behaviour of the production risers.

Tension legs

- fatigue strength
- inspection mode
- the choice of leg design and materials of construction.

Behaviour of the platform under installation

- validity of anchorage placement procedures
- reliability of leg guiding and connection equipment.

Behaviour of the platform under operating conditions

- travel
- stresses caused in the structures.

Behaviour of the production risers

- distortion and constraints in risers
- risk of impact between intermediate risers.

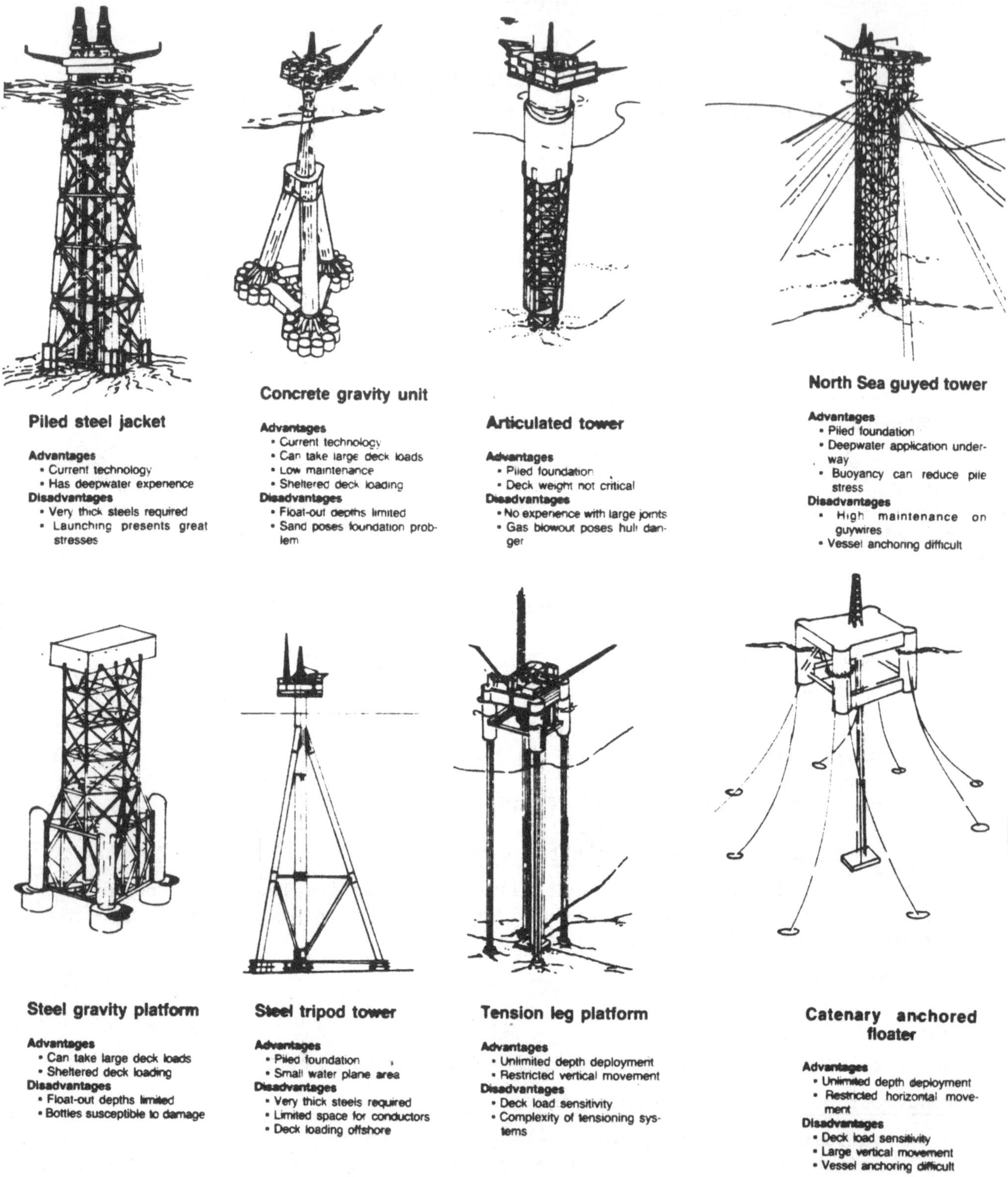

Fig.A.18.1. Proposed platform designs for the northern North Sea.

	Description	Guyed Tower	TLP	FPF
Design Capacity	Payload capability	Light	Heavy	Medium
	Water depth range	600 to 2,000 ft	500 to 3,000 ft	250 to 6,000 ft
	Storm resistance	100-ft waves	100-ft waves	100 ft waves
	Seismic resistance	Poor. Vertical period of vibration approaches 1.0 sec in deep water	Good	Excellent
Structural Performance	Weight vs water depth	Exponential increase above 2,000 ft	Proportional increase with water depth	Proportional increase with water depth
	Motions, lateral	Least motion	Less than FPF	Acceptable
	Motions, vertical	No	No	Yes
	Fatigue, critical areas	Structure	Structure, moorings and risers	Structure and risers
Operations	Safety	Considerable redundancy in mooring system and structure	Moderate redundancy in mooring system and structure. Subsea templates vulnerable to dropped objects.	Moderate redundancy in mooring system and structure. Subsea templates vulnerable to dropped objects.
	Reliability	Reasonable. This concept is based on proved fixed tower technology.	Of concern. Mooring system and subsea system can be relatively complex	Of concern. Subsea system can be relatively complex.
	Maintenance	Difficulty with structure and mooring system	Difficulty with subsea equipment	Difficulty with subsea equipment and mooring system
	Inspection	Difficult	Reasonable	Moorings difficult
	Operability	Conventional. Continuous riser support. Deck level completions.	Tensioned risers. Subsea or deck level completions.	Tensioned risers. Subsea completions. Additional heave motion.
Technology	Engineering	Within current capabilities	Within current capabilities	Within current capabilities
	Fabrication	Small members Uncomplicated	Complex construction Large fabrication facilities required.	Complex construction. Large fabrication facilities required.
	Installation	Considerable time Offshore module installation required	Longer than FPF No modules	Least time No modules
	Potential problem areas	Clump weight lift off Guy line interference Base hinge (if applicable) Spud can foundation	Risers Tension members Tension piles	Risers Deepwater mooring system

Fig.A.18.2. Comparison of design particulars for compliant structures.

PLATFORM TYPE	INSTALLED COST $\times 10^6$	Max No. of wells	Water depth in metres	Well type	Work-over	Oil storage	Reusable
Fixed jacket	350	60	0.200	Land	Yes	No	No
Jack-up (purpose built)	300	30 40	0.150	Land	Yes	Yes	Yes
T.L.P.	250	30	150.600	Subsea-Land	Yes	No	Yes
Semi-submersible (purpose built)	100	15	70.250	Subsea	Yes	(Yes)	Yes
Moored tanker	50	4	50.150	Subsea	No	Yes	Yes
DP tanker	50	1	50.600	Subsea	No	Yes	Yes
Semi-submersible (converted)	30	8	70.200	Subsea	No	No	Yes
Jack-up (converted)	30	12	0.100	Land	No	No	Yes

Fig.A.18.3. Comparison of production platform types.

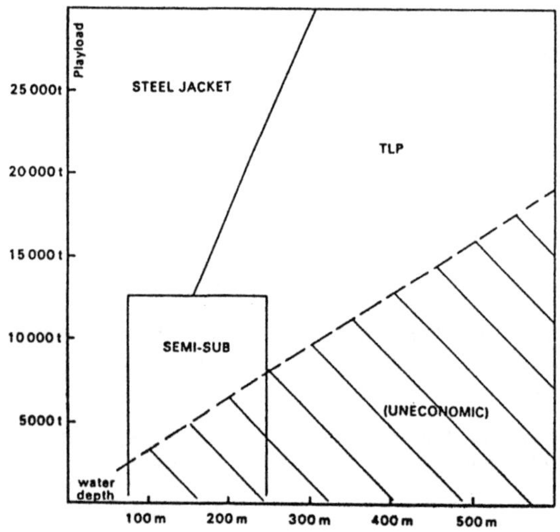

Fig.A.18.4.

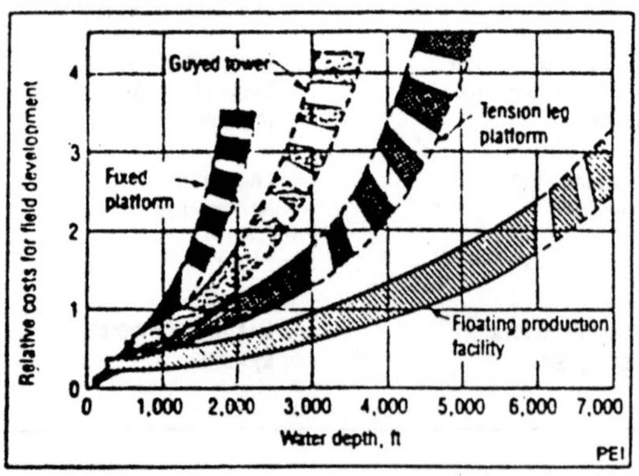

Fig.A.18.5. Cost versus water depth. The relationships shown may be influenced greatly by design environment and payload requirements.

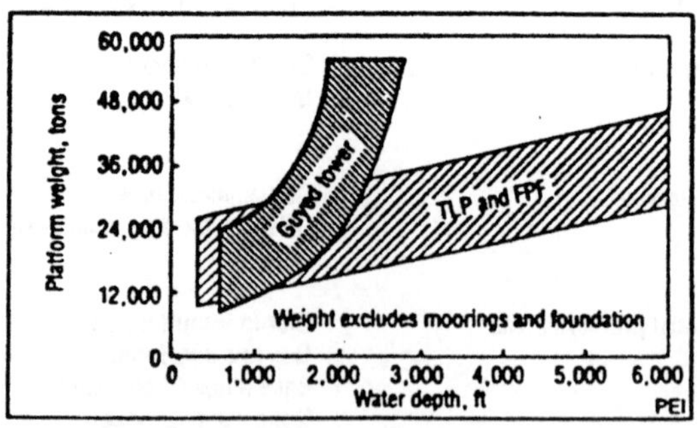

Fig.A.18.6. Weight versus water depth. Weight of the guyed tower increases exponentially with water depth. Water depth has less influence on weight of the TLP and FPF.

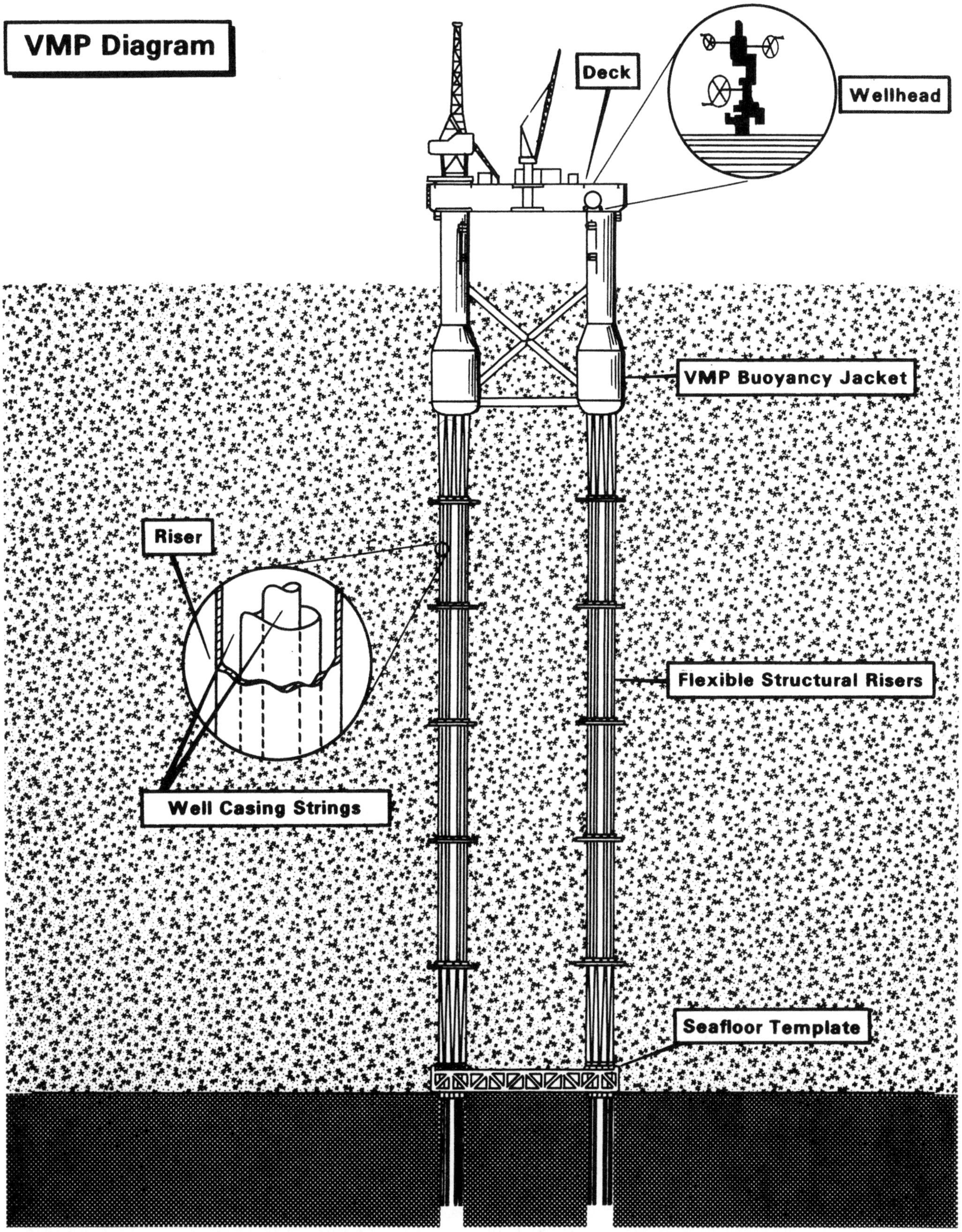

Fig.A.18.7. TLP (Amoco's design).

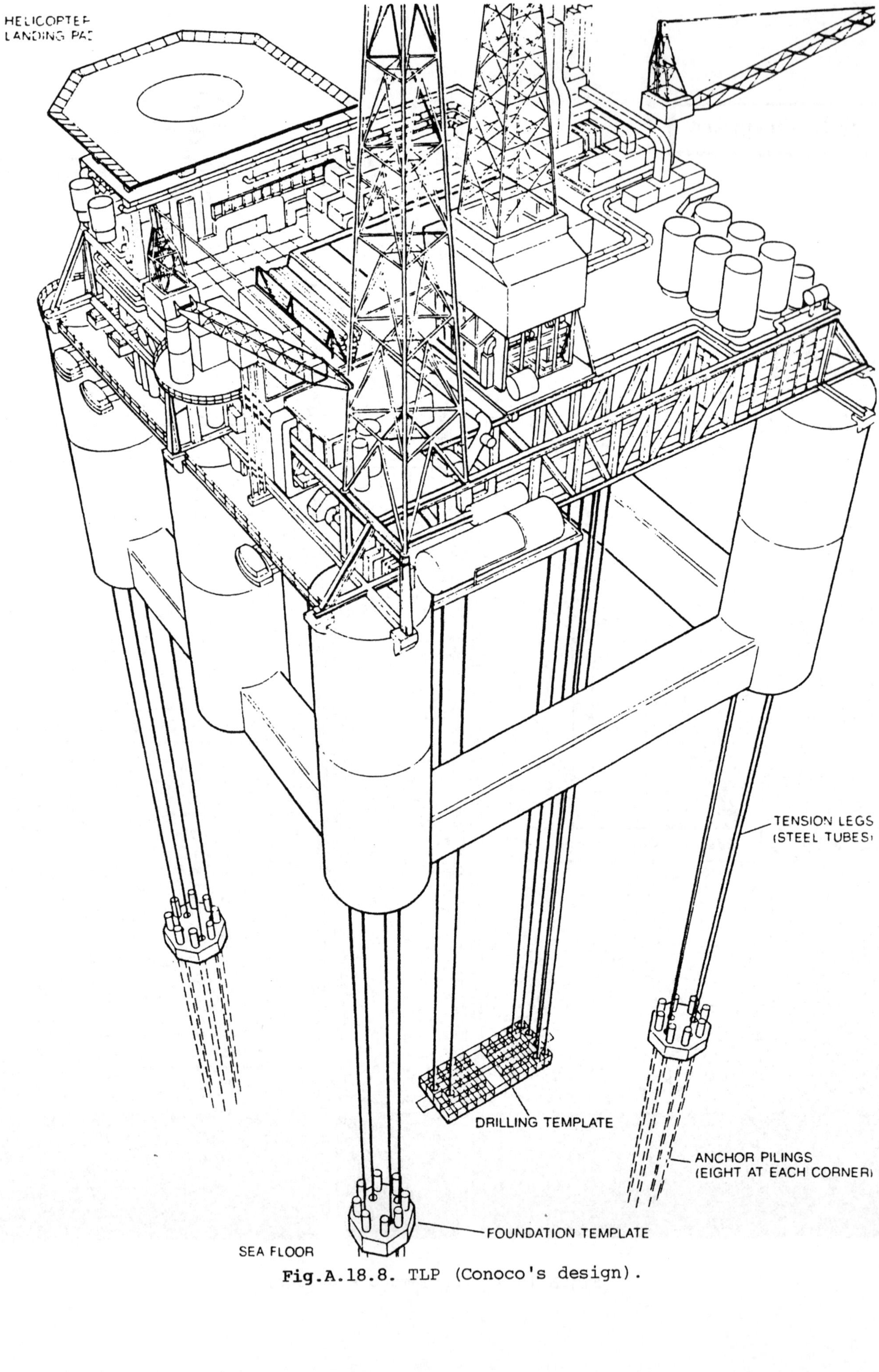

Fig.A.18.8. TLP (Conoco's design).

Appendix 19

Despatch Systems and Offshore Loading Stations

REVIEW OF MAIN COMPONENTS

Theoretically, the same components are found in each of the systems reviewed. As we shall see later, however, the conditions of use are so varied that a number of concepts have had to be developed to fulfill the same function, according to conditions of use.

The essential components are as laid out below.

Fluid Transfer Components

(1) Rotating joints providing for transfer of fluids via the rotary table.
(2) Lines used as risers, providing a connection between the end of the subsea pipeline (PLEM - pipeline and manifold) and the rotating joint.
(3) The loading lines per se, which may be either floating or aerial.

These lines are generally flexible hoses, but in some cases may be articulated rigid pipes.

Mooring Components

These are the following.

(1) 'Loading' buoys, which act as an anchoring point for take-off vessels. These buoys must be self-floating and sea-resistant. In certain cases, the marine environment may lead to replacement of these buoys by articulated columns.
(2) Permanent mooring buoys for the storage ships. In most cases, these buoys are connected to the ship by an articulated yoke, and consideration is even being given to non self-floating systems, as in the case of turret mooring.

The main components of these buoys are:

(a) their anchoring systems, which may be either catenary or tensioned;
(b) their rotary tables.

As above, articulated columns may be used.

The Oil Tanker-Buoy Connecting Arm

The connection between the buoy and storage vessel is via a rigid arm (yoke).

THEORETICAL BASIS OF DESIGN

Regulations

Since these installations are in most cases in international waters, they are firstly required to satisfy current international regulations relating to pollution and safety on board (e.g. SOLAS, MARPOL, and IMCO).

In the particular field of loading buoys, the main classification bodies (Bureau Veritas, Det norske Veritas, American Bureau of Shipping, Lloyds) have their own certification criteria, relating primarily to:

- the design of the buoy and its mooring;
- manufacturing and installation requirements;
- guarantee tests;
- fluid transfer components (in the case of liquid products under transfer conditions).

Engineering and Performance Tests

There are a number of mathematical programmes for calculating stresses due to the sea and wind. Other programmes enable the intrinsic characteristics of floating bodies to be determined, and their movements under these effects to be simulated. These engineering programmes are beginning to replace water tank tests.

Histograms of a ship's movements under the effects of wind, currents and waves, determined either by mathematical programme or by tank testing, enable the dynamic stresses in anchorages and connections between buoys and ships (the hawser or yoke) or between ships to be determined. This movement histogram also provides information on the dynamic stresses to which fluid transfer hoses are exposed.

EXPERIENCE GAINED

Failures of the systems in question are rarely publicised. Nevertheless, certain incidents have been covered by articles in the specialist press. Such incidents are isolated and would not provide sufficient data for a data bank suitable for use in design or preventive maintenance.

Examples

- GAROUPA (Brazil), where the 'CBI' riser in the Campos basin (Brazil) broke through on 7th September 1980. This was attributed to a misunderstanding of the effect of fatigue on the riser shaft under compound bending at its junction with the buoy.
- NILDE (Italy), where one of the articulations in the SALS system arm on the tanker side was damaged in January 1981, a few months after its installation.

From the <u>users</u>' side, a number of users have submitted information on their experiences and the problems encountered. For example:

- ARAMCO, who have reported on 5 years' use of their Jn'Aymah SPM terminal*;

---

* OTC 1981 - Paper 4010.

- SHELL, having tested out the SALS system on Castellion from August 1977 to January 1981 and Tazerka since the end of 1982, and also the yoke tower system on Fulmar in the North Sea, again since 1982;
- SNEA(P), who were the first to try out an SBS system on Ashtart+ in the Mediterranean, and various other systems in the Gulf of Guinea (Kolé, Ekoundou, etc.).

Among other users, particularly of SBS or similar systems, mention might be made of:

- Cities Service, on Poleng (Indonesia) and South Nido (Philippines);
- Atlantic Richfield, on Ardjuna (Indonesia);
- Exxon, on Pulan (Malaysia) and Hondo (Santa Barbara Channel);
- Conoco, on Udang (Indonesia);
- Petrobras, on Garoupa (Brazil);
- AGIP, on Nilde (Italy);
- AMOCO, on Cadlao (Philippines).

On the suppliers' side, particularly those offering complete systems, whether SPM or SBS, a considerable amount of data exists, although it is not publicly available. For example, for 'temporary' type SPM systems:

- SBM
- IMODCO
- SOFEC
- EMH
- Bluewater

and for 'SBS' or 'permanent' type systems:

- SMB
- IMODCO
- Bluewater.

The information available in these companies, most of which are design, engineering and assembly organisations, sub-contracting the manufacture of the sub-assemblies, comes from two sources:

(1) experimental data, from tests of various kinds (wear, fatigue, ageing, etc.) performed on component prototypes (rotating joints, universal joints, etc.);
(2) user data from the compilation of incidents in service (fractures, leaks, etc.).

These various data provide the back-up for the recommendations made by the supplier as part of his after-sales service. Where such data are brought to the attention of the users, we mention this in the brief comments below on the state-of-the-art with respect to main components of the systems.

STATE-OF-THE-ART

Mooring Systems

The most significant types of mooring system are as follows:

- permanent mooring in the North Sea of a 210 000 DWT storage ship on the Fulmar Field (operated by Shell Expro), water depth 82 m;

+ OTC 1976 - Paper 2722.

- permanent mooring in the Mediterranean of a 210 000 DWT storage ship supporting the production treatment installations on the Tazerka Field (operated by Tunirex, a Shell subsidiary), water depth 146 m;
- permanent mooring on the West Coast of the United States of a 50 000 DWT storage ship supporting the production treatment installations on the Hondo Field, water depth 150 m.

It will not be possible to use catenary link mooring lines for greater depths of water or for more difficult seas. Investigations and tests are currently in progress with a view to these future applications (problems associated with the use of steel cables, behaviour of Kevlar cables). In addition, the increase in the capacity of mooring points will require a better understanding of the sea-floors, hence an improvement in the specific means of geotechnical surveying.

Fluid Transfer

At ambient temperature and low pressure, fluid transfer is unproblematical. A number of studies are currently in progress on the transportation of fluids at low temperature (LPG and LNG) and also on the transfer of fluids at high pressure, with respect both to rotating joints and to hoses.

Hinge Pins

On the tanker side, these are generally shells in bronze type alloy, comprising self-lubricating bearings. Although the cause of a major stoppage on Nilde, these hinge pins do not generally appear to require replacement in less than two years.

However, certain suppliers of systems, such as Bluewater, have replaced them with elastic hinges of the SEPFLEX type in bound elastomer, preventing any relative movement between the wearing parts. However, the ageing of this type of material will probably require further investigation.

Rotary Table Roller Journals

These large diameter journals (approximately 3-4 m) do not appear to be a particular source of weakness, the various manufacturers offering a clever doubling system for horizontal rolling faces in case of seizing.

Appendix 20

LPG Sea Loading Stations Using the Single Point Mooring Technique

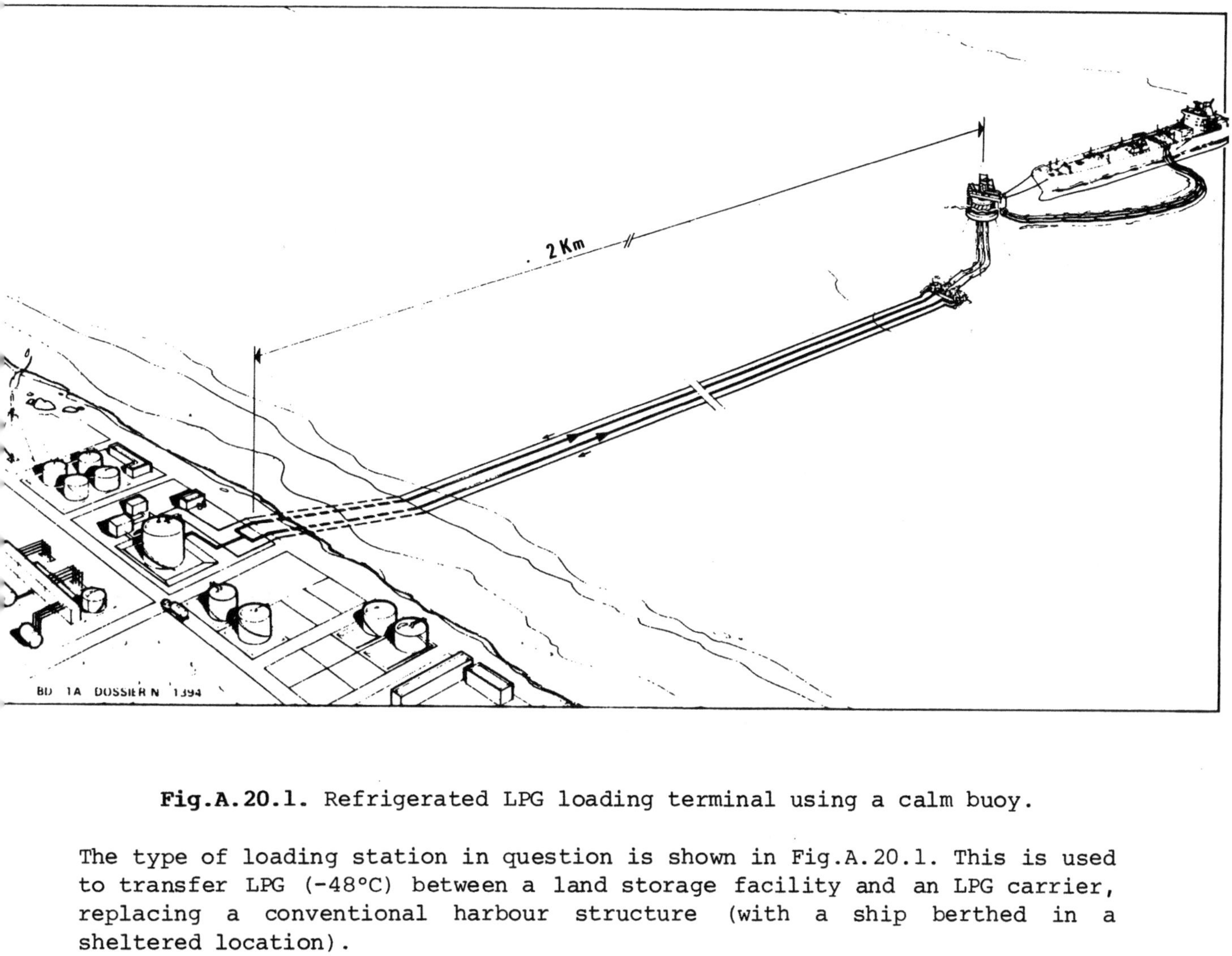

Fig.A.20.1. Refrigerated LPG loading terminal using a calm buoy.

The type of loading station in question is shown in Fig.A.20.1. This is used to transfer LPG (-48°C) between a land storage facility and an LPG carrier, replacing a conventional harbour structure (with a ship berthed in a sheltered location).

The three safety aspects of this system are related to three specific technical aspects:

(1) Cryogenics (-48°C) (materials, shrinkage, resilience, etc.);
(2) Offshore location: at sea, hence exposed to the action of the elements;
(3) Gas: a liquid product which can vaporise as a result of atmospheric heat.

The safety constraints on this type of loading station are dictated by the combination of these three specific aspects.

GENERAL PHILOSOPHY OF THE LOADING STATION

This method removes the loading station from the coast (storage, works and accommodation) (by 2-5 km). Propane tankers are difficult to manoeuvre in heavy winds: the 'aberrations' of the Montoir-Boston terminals, where these ships pass under frequented wells* to moor as close as possible to the works, are avoided.

Major safety aspects in SPMs are the ship's flexibility (still being able to pivot and move in accordance with the elements, even though moored) and the distance from the coast.

In addition, should problems arise, the ship can be quickly disconnected from an SPM and distanced from the loading station, unlike in port, where the ship is tied fast to the quay.

SAFETY ASPECTS BY COMPONENT

(1) Pipe: the pipe must be insulated, hermetically sealed internally for insulation, and in resilient steel. The main risk is fracture from external causes (anchors etc.).

The solution adopted provides maximum safety:

- double concentric steel pipe;
- water propagation stops in the insulation.

(2) SPM: the safety aspect of the SPM is primarily connected to the fact of transferring cold gas. The important points are:

- selection of resilient material for the entire SPM;
- all pipes are unrestricted (expansion joints);
- double containment wall along exposed pipes - maximum safety;
- no valves in the well or body of the buoy - everything above the surface;
- specific valve motorisation design;
- instrumentation, gas flow detection;
- specially designed and tested bi-fluid rotating joint (leak-tightness and insulation).

The various components are thus designed and conceived taking operating safety into account. These components have been especially designed for use in the project, safety requirements having been considered from the outset.

OVERALL SAFETY OF THE TERMINAL IN OPERATION

The operating safety requirements are as follows:

- the LPG lines must be kept constantly cold (underwater pipes) between storage and the PLEM;
- a compulsory procedure must be established for purging hoses and SPM systems and supplying with inert gas after each loading operation;
- a valving system must be provided for emergency conditions to enable appropriate sections of LPG lines to be isolated and hoses to be isolated by two valves. All safety valves must be motorised.

---

* Translator's Note (Sic).

These requirements were taken into account in designing the SPM and SLEM. The safety assessment (risk tree) and procedures for use of the terminal were to be completed by the end of 1983.

Appendix 21

Safety of Subsea Pipelines

GENERAL

The past ten years has seen a constant increase in the number of subsea pipelines laid and commissioned. This growth has been accompanied by a parallel increase in the depth of water in which such pipelines have been placed, with increasingly severe environmental conditions.

While a subsea line is basically no different from a land line in its design and structure, the costs of construction, operation, maintenance and repairs are considerably higher. The importance of reliability in subsea lines is therefore evident. To achieve maximum reliability, it is necessary to be duly familiar with the various risks run by sea-lines, in order to define the line inspection and maintenance schedules and the resources required to implement these schedules.

The object of this section is to review the risks liable to affect pipeline safety, and then briefly to describe the possible methods of protection. We shall then give a brief description of the methods of inspection commonly used at the present time. With regard to existing literature on this subject, unfortunately the safety of subsea lines has so far failed to raise as much interest as the safety of mobile or fixed offshore structures which are, after all, inhabited by man, and literature is far from plentiful.

POTENTIAL RISKS

Description of the Risks

Risks for subsea lines may be divided into two main categories: natural risks and risks due to human activities. Table A.21.1. summarises these risks; comments on the main risks are given below.

Table A.21.1. Potential risks for subsea lines.

Natural risks		**Risks due to human activities**		
Environment	**Natural and exceptional phenomena**	**Offshore activities**	**Operation**	**Deficiencies in pipeline installed**
External corrosion	Cyclones - storms	Shipping accidents	Operating errors	Design errors
Waves	Earthquakes	Dragging anchors	Equipment malfunction	Concealed faults in construction
Currents	Sediment transportation	Fishing activities	Internal corrosion	Damage caused in burying
Tides	Erosion - scouring	Dredging	Sabotage	Equipment deficiencies
Wind	Floor phenomena	Dumping of rubbish	Explosion	Lining damage caused in laying
Marine organisms	Bed fluidisation		Fire	

Natural Risks

(1) Environment. The risks resulting from the environment are essentially due to:

(a) environmental aggressivity, exhibited by external corrosion of the pipeline material;
(b) hydrodynamic effects of the waves and currents liable to affect the stability of lines, whether buried or unburied.

The problem of marine organisms must be considered, in particular on the vertical parts of lines, at platform risers: the weight of living organisms attaching themselves to the pipes can cause dangerous loadings.

(2) Natural and exceptional phenomena. These phenomena may be classified into two groups:

(a) Accidental phenomena of limited duration:
- cyclones and severe storms;
- earthquakes;
- underwater landslides.

These phenomena are always violent and frequently highly damaging to subsea lines.

(b) Permanent or continuous phenomena. These relate to sediment transportation, erosion, and scouring. They have a number of effects:
- uncovering buried pipes;
- creating free spans, i.e. portions of lines no longer resting on the floor, as a result of scouring.

These free spans may then cause inadmissible mechanical bending stresses and vibration phenomena (vortex shedding) due to transverse currents.

The risks of free spans is most critical in the short term.

Risks Due to Human Activities

(1) Risks deriving from offshore activities. The two main risks are:

(a) Dragging of the pipeline by ships' anchors. The risk level is of course dependent on a number of factors:

- the depth of water;
- the size of anchors;
- the pipeline diameter;
- pipeline protection (i) whether or not buried, and (ii) the presence and quality of concrete cladding.

(b) Fishing activities.

The major risk is dredging of the pipeline and impact caused by trawls. Risks due to anchors are more frequent at the edges of platforms, at construction and service vessel moorings, than in the general sections of sea-lines, where the risks of trawl dredging are greater.

Regulations on navigation and mooring in these zones are designed to minimise these risks, but they are not always observed particularly in cases of emergency.

Similarly, damage caused by the accidental deposit of rubbish or other items is generally localised around the edges of platforms.

The consequences of these various types of aggression extend from loss or damage of the concrete ballast cladding and corrosion-proof cladding to complete fracture of the line.

(2) Risks deriving from operation. The risks associated with operation and maintenance are primarily due to errors in manoeuvring associated with malfunction of the safety devices.

This type of incident is more frequent at the time of commissioning the pipeline, and the consequences are not generally catastrophic.

In spite of the safety precautions taken, fire and explosion risks can never be zero, since no safety device can attain 100% reliability.

Risks of sabotage do exist, but in our view they are relatively low, given the substantial nature of the diving resources required to mount such an operation. Offshore platforms or onshore terminals are more vulnerable targets.

(3) Deficiencies in the installed pipeline. When an engineering company with considerable experience in subsea lines is used, design error remains an extremely low probability. In most cases, incidents are due mainly to inadequate or defective inspection on acceptance of the materials and equipment, or during construction.

It should be noted that the majority of these defects, if not revealed during construction, will nevertheless be identified at the time of hydrotesting and inspection undertaken prior to commissioning.

Level of Damage

The classification below endeavours to establish a hierarchy in the risks run by subsea lines in terms of economy and environmental attack.

Risks Causing Major Damage

- External corrosion - location of leaks which may occur as a result, and the necessary repair work, are difficult and costly.

- Cyclones and heavy storms - can result in total destruction.
- Earthquakes - can fracture the lines.
- Shipping accidents and dragging anchors - damage which can be as serious as fracture.
- Fishing activities - as above.
- Internal corrosion - this progresses slowly until the weakened pipe eventually fails under the effect of external forces.
- Sabotage, explosion and fire.

Risks Causing Moderate Damage

- Erosion, scouring, providing the effect of these phenomena is corrected before they become too great.
- Waves and currents, provided their force does not greatly exceed the force specified or estimated at the design stage.
- Accidental dropping of rubbish and other objects.
- Minor hidden construction defects. Defective welds would clearly not come under this category.

Risks Causing Minor Damage

- Marine organisms.
- Operating errors, providing safety systems are functioning correctly.
- Design errors - these are minor errors, as engineering is assumed to comply with good practice.

Probability of Occurrence

It is useful to have information on the relative probability of occurrence of the various risks. These risks have been classified into three categories:

Highly Probable

- External corrosion.
- Dragging anchors.

Probable

- Erosion, scouring and floor phenomena.
- Operating errors.
- Equipment malfunction.
- Hidden construction defects.
- Internal corrosion.

Improbable

- Earthquakes.
- Cyclones and heavy storms.
- Sea conditions exceeding the design criteria.
- Sabotage, explosion and fire.
- Risks associated with fishing - lines are generally buried or protected in fishing zones.
- Design errors.
- Material defect.

METHODS OF PROTECTION

General

To preserve the pipeline against the majority of the risks listed above, or to attenuate their effects, the most logical solution is to protect it against the direct action of the elements. There are two methods commonly used: burying or placement in a trench, and covering with various materials.

Burying

Burying is generally carried out during the pipeline construction phase, as the technique can be dangerous to perform in operation.

Normally, partial protection only is provided, as the pipeline remains uncovered although below the mean level of the floor, a trench being dug by mechanical means.

When seabed conditions permit, the bed may also be fluidised by means of pressurised water jets. The pipe then sinks under its own weight into the fluidised sea bed.

Cover

When the nature of the seabed does not allow trenching to be carried out under reasonable technical or economic conditions, the pipe may be covered. Various means and materials are used:

- juxtaposed reinforced concrete shells;
- anti-scouring cushions;
- materials poured from the surface, either directly or through a guide pipe, which gives greater accuracy of placement and allows material savings (the VELPO system).

Reduction of Free Spans

Excessive free span lengths may be reduced by installing pipe supports. These supports take the form of pyramids of sand bags, or cement mortar may be injected from the surface by means of hoses, into flexible jackets (rubber or plastic) between the sea-floor and the pipe. Cushions may also be arranged on top of the pipeline, to give stability.

CONCLUSIONS AND RECOMMENDATIONS

Risks

The following assumptions apply.

(a) The basis of design for the pipeline, with particular reference to data relating to meteorological and sea conditions and sea-floor characteristics, are sufficiently accurate and conservative to allow correct sizing of the pipeline with respect to:

- steel grade and thickness;
- appropriate corrosion proofing;
- suitable concrete ballast.

(b) The pipeline has been built, installed, tested and commissioned according to good engineering practice, has suffered no major damage, and exhibits no hidden defects.

Under these conditions, it may reasonably be assumed that the majority of the potential risks have been eliminated or that their consequences on the line throughout its service life have at least been greatly reduced.

What major risks remain? In our opinion, there are two.

(1) Corrosion, and particularly external corrosion, for a number of reasons, including:

- gradual deterioration or accidental damage to corrosion-proofing;
- defective cathodic protection, loss of anodes, excessive consumption, etc.

(2) Mechanical stresses in the pipe due to movements of marine floor sediment, erosion and scouring, leading to modification of the sea-floor topography. The major resultant risk is the creation of free spans, i.e. unsupported pipe runs.

Subsea Inspection

Pipeline inspection is essential, both directly after placement, to ensure that the work has been carried out correctly, and in service, to forestall any serious damage and identify maintenance and repair work required.

Inspection methods are complementary, hence no one method is preferred.

- Human intervention, the use of divers or manned submersibles with suitable equipment will in many cases continue to be irreplaceable.
- Acoustic inspection is an efficient, relatively low-cost and fast method for identifying areas in which a more detailed specific inspection is required.
- Specialised remote controlled units capable of dealing with various problems simultaneously are also an indispensable method of inspection. Being relatively unaffected by depth, this method is destined for major development in the future.
- Internal inspection by special-purpose pigs is an easy, low-cost method: although its use is still limited, development possibilities are considerable, with a potential for moving into new areas.

Future Developments

In the light of the comments in the preceding section, operators of major subsea pipelines, where the economic effects of shutdown are considerable, will greatly welcome any improvement to existing inspection techniques. Improvements must relate to reliability, measuring accuracy, and the ease and speed of use.

In our view, there are two particular areas of interest:

(1) multi-purpose remote controlled units; and

(2) instrument pigs capable of continuous recording of a large number of parameters.

Appendix 22

Sounding Possibilities for Underwater Structures in Concrete

INTRODUCTION

Non-destructive sounding techniques are widely used on concrete structures on land. Sounding is perfectly feasible for underwater structures, but the possibilities offered by conventional methods are limited, and the interpretation of measurements has to take into account the mechanical behaviour of concrete impregnated with water.

Impregnation with water modifies the speed of propagation of sound waves and when strength tests are required, it is essential to carry out calibration on speeds of sound and crushing strengths on saturated underwater concrete. Cracks fill with water and will not, therefore, give a major discontinuity in the sound impedance in the medium. The sonic method can no longer be used, and methodologies involving electrical methods must be employed. The method developed by the CEBTP applies to reinforced concrete, and consists of detecting cracks in reinforced concrete, highlighting lines at which preferred current circulation can occur.

Contractors and operators are always concerned to identify potential developments in structures liable to prevent such structures from fulfilling their design function. Structural inspection and surveying is designed to detect such developments, and to provide operators with the necessary information on which to take maintenance decisions. Information is obtained by examination, based on the results of instrument measurements, determining variations in states of stress or distortion, and by sounding, aimed at ascertaining the state of health of the structures, analysing the propagation of sound waves, particularly in the case of concrete structures.

The sounding of concrete structures in air is a conventional and easy technique for obtaining information on the mechanical strength of concretes and their homogeneity, and measuring the depth of any cracks.

In the case of underwater structures, however, subject to particular stresses (the action of sea-water on cement and steels, fatigue due to the cyclical action of the waves) the possibilities offered by conventional methods are limited, since impregnation by water makes it impossible to investigate the depth of cracks. Being filled with water, these cracks do not give sufficient variation in the sound impedance to be detected.

Within the framework of research programmes under the patronage of ARBEM, the Association de Recherche sur le Béton à la Mer (the Offshore Concrete Research Association), the CEBTP, in conjunction with CNEXO, has studied the application of dynamic sounding to underwater structures and has developed a crack detection methodology for reinforced concrete structures using electrical methods.

DYNAMIC SOUNDING OF UNDERWATER CONCRETE STRUCTURES

The speed of propagation of longitudinal sound waves within a medium is governed by a classical law involving density, dynamic modulus of elasticity, and the Poisson coefficient.

What the contractor or operator wants to know is the mechanical strength, or more specifically the crushing strength, of the concrete. To evaluate this parameter, other factors such as the mineralogical nature of the constituents of the concrete, the state of hydration of the binder, etc. have to be used, and only by experimentation will it be possible to ascertain the empirical relationships between the crushing strength of the concrete (Rc) and the speed of propagation of the waves (Vs). A number of formulae have been proposed: these are parametric formulae, which may for example, be:

$$Rc = Ae^{BVs}$$

A and B being parametric coefficients which have to be determined empirically.

Structures are therefore sounded, either using known parameters (if sufficient identifying information is available) or (and in this case the method is much more precise and more rigorous) two or three speed of sound measurements are carried out on specimens taken from the structure by core sampling, these being subsequently crushed.

The crushing strength of a concrete is extremely important, since it governs its mechanical behaviour, but all other things being equal, it also enables the compactness and durability of structures to be assessed. In any cases of doubt, therefore, the constructor or operator will always want to ascertain the mechanical strength of the concrete in situ. In the case of offshore structures, an understanding of this parameter is even more important, given the risk of pathological action of the environment on the cement.

What Effect does Immersion have on the Properties of Concrete?

An experimental investigation designed to determine the influence of the depth of immersion on the strengths of concrete and the propagation of sound waves has been carried out on three types of concrete:

(1) Low quality - Cement content 200 kg/m^3; water/cement ratio = 1; 28 day crushing strength (Rc 28) = 12 MPa.

(2) Average quality - Cement content 320 kg; water/cement ratio = 0.6; Rc 28 = 23 MPa.

(3) High quality - Cement content 400 kg; water/cement ratio = 0.48; Rc 28 = 43 MPa.

At two months, some specimens were simply immersed, while others were impregnated with water at a pressure of 15 bars, then immersed. They were then monitored for a year, and compared with control specimens kept in air. The conclusions of these investigations were as follows.

(1) Concretes become progressively impregnated, their water content approaching saturation point.

(2) Immersed or impregnated concretes exhibit totally comparable strength patterns. When compared with concretes in air, they exhibited an increase in crushing strength which was all the more marked, the higher the initial

mechanical characteristics (increase in strength of 11% for concrete A, 29% for concrete B, 21% for concrete C).

(3) The speed of sound is affected by the presence of water. Immersion involves a considerable increase in the speed of propagation of sound waves. The increase in this parameter is not solely associated with the increase in strength: it is largely due to saturation, and for this reason it is much greater in low quality concrete with a high level of porosity than in high quality concrete with a low level of porosity. It is substantially greater for impregnated concrete than for immersed concrete. Increases in the speed of sound for immersed and impregnated concrete respectively are:

- 8.83% and 9.45% for concrete A;
- 4.5% and 5.45% for concrete B;
- 1.57% and 2.90% for concrete C.

The variation in the speed of propagation of longitudinal sound waves may seem relatively low, but in the light of the laws giving crushing strength as a function of speed of sound, these will be seen to relate to very major variations in crushing strength (see Fig.A.21.1.). It is not therefore possible, without considerable risk of error, to extrapolate the values of the parameters from the ratios of crushing strengths to speed of sound used for sounding structures in air. For control purposes, a special calibration exercise should be carried out for immersed concretes. Contractors building offshore structures might also, and quite logically, be advised to perform long-term crushing strength and speed of sound tests on immersed specimens of their design concretes: this would then provide them with the necessary parameters from which to interpret soundings. It should be mentioned here that French contractors have been duly involved in the work being performed by ARBEM.

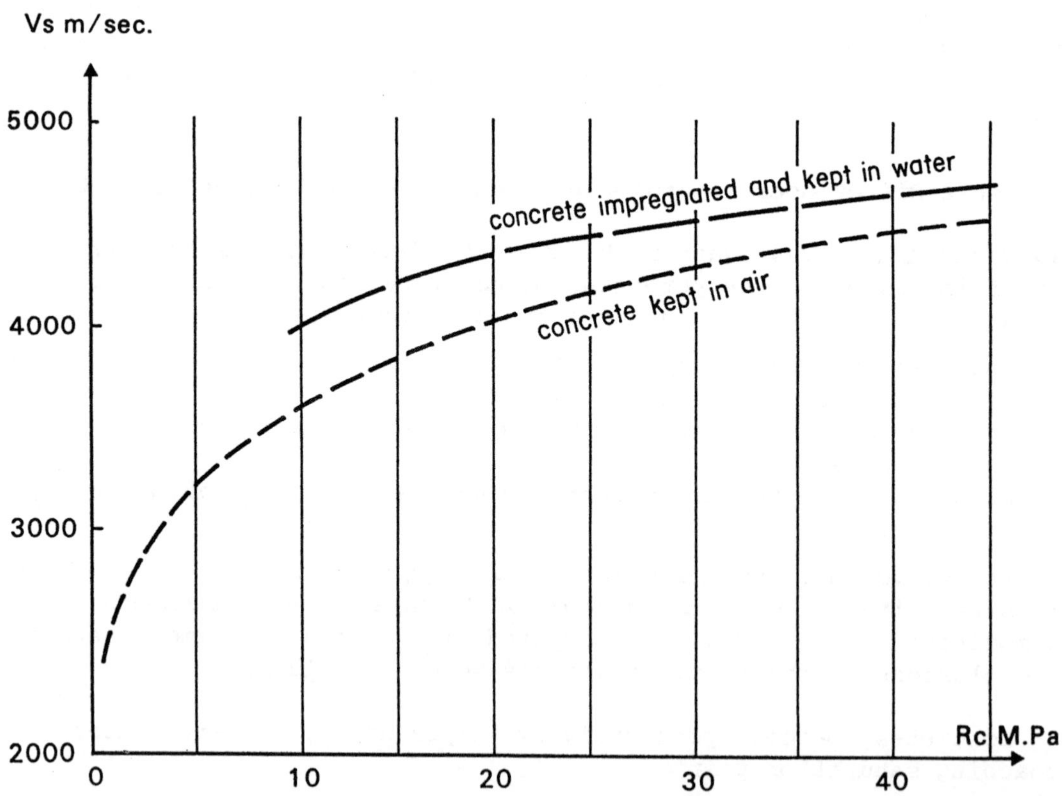

Fig.A.21.1. Relationship between the speed of sound Vs and crushing strength Rc

We shall not, here, analyse these factors in greater detail. It is enough to say that these phenomena are perfectly logical and foreseeable from a knowledge of physical chemistry and the mechanics of hydraulic concretes.

What Conclusions Can Be Drawn?

The dynamic sounding of concrete in underwater structures thus allows the mechanical characteristics of concretes to be assessed in situ, i.e. it can indicate whether or not any pathological or mechanical phenomena have produced internal developments or disaggregations within the concrete.

On a practical level, there are no particular problems in testing. Using conventional measuring equipment connected to conductors and pressure-tight transducers, tests can be carried out by a team of divers connected by telephone to the technician following and directing operations from the surface, collecting data from the detecting and analysing instruments.

Experiments carried out at the COB pool have confirmed the possibility of using conventional methods of sounding, i.e.:

- direct transparency: measuring the time taken for sound waves to propagate through a wall or component;
- on the surface: measuring the time taken for waves to propagate between a fixed transmitter and a receiver, moved at a constant distance, and used to deduce the speed of sound in the material in question.

The problem of calibration may be treated in the same way as on structures in air, using known parameters, if available, or carrying out a number of core sampling operations in the underwater concrete.

CRACK DETECTION

On structures in air, cracks can generally be detected quite easily by visual inspection, dynamic sounding being used to confirm in cases of doubt and to determine the depth of cracks, studying the discontinuity of transmission.

In an underwater medium, structures are frequently covered by exogenous items (algae, shells, etc.) and inspection becomes a more delicate operation. Cracking here is a very important factor, since the saline medium is aggressive to reinforcement, corrosion of the reinforcement causing cracking, and since fatigue phenomena cased by the cyclical action of the sea, can also cause the same effects (i.e. cracks).

As sonic methods cannot be used, the CEBTP have developed an electrical method; its principle is detailed below.

A difference in alternating potential (frequency 100-10 000 Hz) is established between the reinforcements and a special flat conductor target, held parellel to the surface, as close as possible.

If there is no crack penetrating to the reinforcement, current is circulated by conduction of the concrete and the current density perpendicular to the wall is more or less uniform. In the presence of a crack penetrating to the reinforcement, a major increase in current density is observed showing an extremely marked discontinuity at the level of the concrete, but tending to become less distinct on moving away from the surface.

An offshore structure may have a surface area of many m^2, and in order to be practicable, a method such as this must provide for relatively rapid examination of a large area. Systematic scanning is therefore required, and the device developed by M. Paquet at the CEBTP combines electronic horizontal exploration over as wide a band as possible, of one or more metres, and vertical exploration, obtained by progressive mechanical climb of the measuring device up the wall.

The analyser system should provide an image of the current density distribution over the area under investigation.

The method used consists of making up a target of discontinuous components applied to a potential V, and measuring the current crossing each component of the target, successively switching these on a single conductor. Horizontal scanning can be performed at a frequency corresponding to a preset vertical movement of 10, 20, 50 mm, etc.

In practical terms, the device outlined in Fig.A.22.2. may be envisaged. This comprises an insulating panel, its internal face covered with a conducting plate, cut at the centre of which are aligned electrodes, moved over the panel by an operated mechanical device (e.g., a winch and cable), on the surface of the concrete.

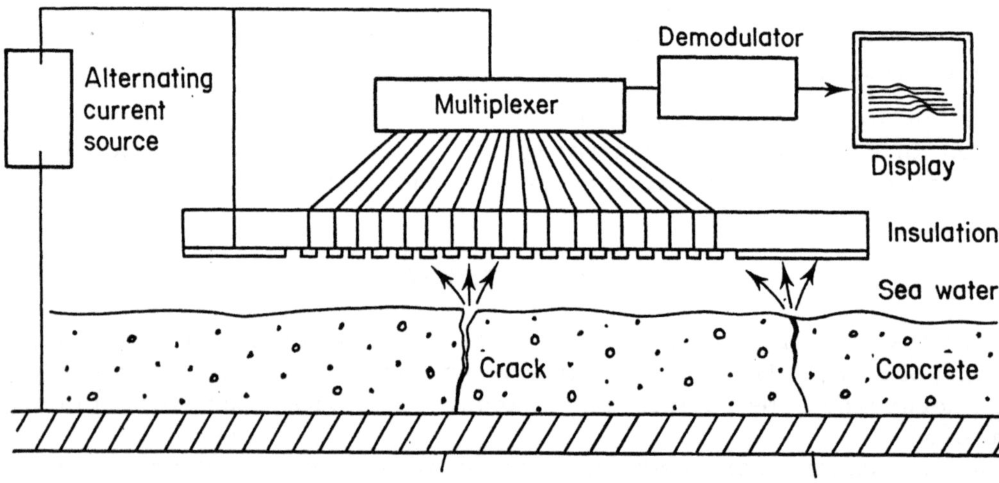

Fig.A.22.2. Electrical crack detection. Diagrammatic representation of the measuring system.

An alternating current is applied between the metal reinforcements of the reinforced concrete structure and the conducting plate. Under the effect of the potential difference, currents are generated between the reinforcement, the electrodes and the plate: the current is low if the concrete is sound, and exhibits a high local value in the vicinity of any crack. Analysis can be carried out electronically by an analogue multiplexer, successively switching the current from the electrodes to a demodulator, giving a voltage that can be displayed on an oscilloscope, doubled with a recorder. Vertical scanning of the oscilloscope may be controlled with the vertical displacement of the panel and an image will thus be obtained of the surface of the concrete on which any network of cracks will be revealed by the concentration of current lines.

CONCLUSIONS

The development of offshore structures over the past decade has led researchers to consider the behaviour of concretes in underwater structures and means of sounding these structures. To this end CNEXO and CEBTP have since 1976 combined efforts to cooperate on joint research programmes, and in 1979 this exercise was expanded within the framework of ARBEM.

Research programmes into the impregnation, cracking, fatigue and sounding of concretes were thus initiated, and we felt that it would be of interest to this symposium to present certain practical results on the sounding of structures.

Dynamic sounding could become a conventional operation, providing information on the mechanical characteristics of concretes, i.e. their development in time. Existing equipment could be easily adapted and brought to an operational status.

The action of sea-water on reinforced concrete (a subject on which the author has had occasion to present a paper to the International Conference on the Behaviour of Concrete Structures Offshore at Brest in October 1980) begins by generating cracks, and it could be of advantage for these to be detected as early as possible.

The CEBTP has accordingly carried out laboratory tests on a highly promising electrical method which will undoubtedly be of great interest to specialists in offshore inspection.

Appendix 23

The Hyperbaric Welding Method of Pipe Connection

THE PRINCIPLES AND PROCEDURES OF HYPERBARIC WELDING

Hyperbaric welding consists of creating a gaseous atmosphere at the required weld, the ambient pressure being equal to the hydrostatic pressure of the place of work, i.e. 15 bars for a 150 m depth. This dry atmosphere is produced inside a welding chamber, itself built inside a pipe alignment frame.

The quality of the atmosphere is permanently monitored and regenerated throughout the operation, on the one hand to maintain a breathable atmosphere, and on the other to provide satisfactory chemical exchange at the actual welding arc.

The procedure followed for connection by hyperbaric welding is as follows:

(1) With the pipes close together, the concrete is removed from the two pipe ends by performing a longitudinal and circumferential cut using a hydraulic circular saw.
(2) The corrosion-proofing is then removed, to cut off surplus lengths of pipe by either oxyacetylene or mechanical cutting.
(3) Pipes are then pre-aligned with the aid of lifters or H-frames after inserting a plug in each (the purpose of this plug is described later).
(4) The welding chamber pipe alignment frame assembly is then lowered across the required connection using the surface support crane and guidelines, to achieve accurate positioning on the sea-floor. The guidelines are fixed to each pipe end.
(5) Using the lifting systems already placed in position on the pipes during the preliminary phases of concrete removal and pre-alignment, the ends of the pipes are positioned in the alignment frame grippers.
(6) The welding frame is then hermetically sealed around the pipes by means of doors which are threaded onto each end, each door comprising a flexible hermetically sealing skirt, tightened around the pipe.
(7) The plug inflation is then adjusted, finally allowing the supply of gas to the welding chamber, to displace the water contained.
(8) Pipe fitting and welding personnel are then moved into the welding chamber.
(9) The pipe fitters then carry out final alignment of the pipes using the alignment frame, hydraulically operated from inside the welding chamber.
(10) Fine pipe fitting adjustment is carried out by mechanically cutting off any surplus lengths and chamfering the ends.
(11) The connecting component, known as a sleeve, is positioned between the pipes, and welding per se can then begin.
(12) After making the two weld seams, a gamma radiography check is carried out to ensure that the welds are of the required quality.
(13) The corrosion-proofing is then put into position before refilling the chamber with water for recovery at the surface.

This procedure using an alignment frame assisted by a lifting device has been specially developed to reduce the stress on the alignment system, hence its weight and cost.

In effect, the use of a lifter enables an increase to be made in the length of pipe used in alignment, thereby reducing the moment of alignment required of the alignment frame. This technique enables connections to be made using equipment of not more than 45 tonnes in weight, as compared with the 70-80 tonnes for the conventional equipment.

HYPERBARIC WELDING EQUIPMENT

In addition to the conventional saturation diving equipment, hyperbaric welding involves the use of the following items:

- two lifters or H frames;
- a Sea Horse alignment frame;
- a subsea welding chamber;
- a subsea welding assistance module.

Lifter

A lifter, using hydraulic power, is an underwater lifting and shifting crane for pipes with outside diameters of up to 42 inches. It comprises two vertical posts of square section, serving as a guide for the vertical movement of the horizontal beam. This beam supports a horizontally moving trolley. The pipe is secured in the trolley by a set of belts, held by two hydraulic jacks. These devices have been used intensively in the production of hyperbaric weld connections in the North Sea. Ten are now in operation.

For some applications (e.g., surface recovery on completion of the job) the following movements may be carried out without an umbilical:

- pipe descent to the floor;
- belt removal;
- the trolley can be locked in position or can remain free for the shifting operations.

When the pipe is locked in the trolley, the lifter is able to move laterally to its new operating position. This is particularly useful when a lateral displacement of more than 2.5 m is required.

Pipe Alignment Frame (Sea Horse)

The Sea Horse pipe alignment frame essentially comprises:

- a central beam secured to the top of the welding frame;
- a clamp at each end of the beam;
- a control panel inside the welding chamber.

The clamps, moved horizontally, consist of:

- a universal joint on the welding chamber side;
- two gripper opening and closing actuators;
- two orientation mechanisms.

The alignment frame is supplied from a hydraulic unit on the surface, by means of an umbilical. It can be operated by a diver in the water or by pipe fitters/welders inside the welding chamber. The alignment frame structure is independent of the depth of water, as none of the compartments are hermetically sealed.

The Welding Chamber

The permanent equipment units in the welding chamber are limited to:

- internal/external lighting;
- gas supply;
- tool clamp control panel;
- TV camera;
- handling devices;
- hot water heating system.

The assistance module receptacle is clamped to the side of the welding chamber. Two lateral openings enable the pipes to be introduced into the chamber. When the compartment is positioned above the pipes, and once the clamps have been locked around the pipe, the side doors are positioned around the ends of the pipes and locked onto the chamber. Hermetic sealing is then achieved by means of rubber sleeves.

There are three doors mounted onto the chamber:

(1) an internal door for dry transfer;
(2) an internal side door giving access to the assistance module;
(3) an external diver access side door.

A hydraulic jack is provided at each corner of the chamber for level adjustment on the sea-floor. Each jack is fitted with a swivel joint and a circular foot. The jacks are individually controlled from inside the chamber.

The Assistance Module

In view of their experience in hyperbaric welding and the evolving market, COMEX have developed a system based on known and tested components - the assistance module.

First generation welding chambers had in-built equipment: gas recycling, welding sets, pre-heating panels, gas and communication panels.

In addition to the space in the welding chamber required for these items, they have also sometimes been the source of operational problems in cases of accidental disturbance.

The idea of the assistance module is to separate the various functions: (1) the welding chamber - a bare chamber containing only a minimum of equipment, and the assistance module receptacle, and (2) an independent module containing the equipment, connected to the welding chamber. This concept has four major advantages.

(1) The interior of the chamber is kept clear, giving improved access around the pipe.
(2) The compartment is extremely simple, providing the possibility of adaptation to any particular type of fitting.
(3) Fragile equipment is easy to recover (in the event of site abandonment for reasons of the weather conditions).
(4) Possibility of repairing without major interference in operation (a spare 'MISS' can be kept on board if required).

The system comprises the following component parts.

(1) The module itself, which is clamped onto the welding chamber.
(2) The module receptacle, secured to the side of the welding chamber.
(3) A system for handling the module without involving the surface diver.
(4) The power unit.
(5) The welding inspection unit, containing:
- remote controls, communications, and gas controls;
- the carrier umbilical permanently connected to the module.

The module contains all equipment items customarily incorporated in the old welding chambers (welding sets, atmospheric recycling, preheating, grinding wheel supply, gas supply, etc.).

A door accessible from the inside of the chamber provides access to the module after clamping the latter to the receptacle from the inside of the chamber and draining the clamping lock by means of the umbilical, and purging with gas.

HYPERBARIC WELDING

Problems Involved in Hyperbaric Welding

There are two main difficulties: (1) physiological, and (2) technical.

The use of welders working under pressure conditions exceeding atmospheric pressure is subject to certain diving rules. Extremely stringent rules have to be followed to avoid the high pressure nervous syndrome (HPNS). This syndrome may occur past a certain depth, and is accompanied by a marked reduction in working capability on the floor. The speed of decompression is also important, directly affecting the state of health of divers during and after the saturation working phase.

Provided these diving rules are strictly adhered to, there is no fear of physiological problems. The parameters to be monitored are: (1) oxygen, carbon monoxide, and carbon dioxide, (2) ambient temperature, and (3) relative humidity. The values required for these parameters are a function of pressure, and can create difficulties with respect to the parameters required for welding.

From these comments, it is clear that particular processes and procedures have to be developed, paying special attention to possible variations in the environment of the arc and metallurgical exchange under pressure.

The particular technical difficulties involved in hyperbaric welding lie in the significant variations in voltage, the intensity of the arc under pressure and the metallurgy of the weld. The quality of the result is also greatly dependent on the welding process used. The general trend is towards an increase in arc voltage due to constriction of the pressure arc column. An increase of 5-10% in the case of MIG welding is observed for pressures of around 15 bars.

In addition, the stability of the arc and the regularity of transfer of metals tend to fall as the pressure increases. This instability is primarily a function of excessive heat loss at the arc and melting bath producing low arc penetration.

These effects may, however, be efficiently countered by improving the ionisation of the arc by the filler products and by adapting current sources to the critical requirements of the welding arc. The type of welding set and its dynamic characteristics have a major influence on the quality of transfer and also depend on the combination of wire/gas used. The arc is more stable if a flux cored wire is used as opposed to a solid wire, but positional welding then becomes more difficult.

Dense smoke results in low visibility, particularly in electrode welding as opposed to MIG welding, and also particularly in the penetration pass, where the smoke is more difficult to draw off.

An atmospheric recycling system allows the dusts in the smoke to be filtered and a breathable atmosphere to be produced in the welding chamber. When the arc is ignited, the calcium carbonate contained in the basic electrodes decomposes. The reaction is generally incomplete, and the formation of carbon monoxide is observed, breaking down into soluble carbon in iron. Other oxide reactions (reduction or desulphurisation) also differ when occurring under pressure.

The mechanical characteristics are also changed - particularly hardness and elongation.

Another important factor is the dissolution of hydrogen in the fused metal, at its greatest under pressure. This effect can involve more significant cracking in the heat affected zone than at atmospheric pressure.

Hyperbaric Welding Processes

These processes are TIG, MIG and electrode (MMA) welding.

TIG (non-consumable electrode and wire filler)

Advantages	Disadvantages
All positions	Special 400 A water cooled torch
All materials	Thickness limit
Self-cleaning arc	Low rate of deposition
No flux	Cost of protective gas
Concentrated heat source	Tungsten inclusions
Calm melting bath	Electrode behaviour under pressure
Low electrode wear	Pressure-sensitive arc
Little smoke	Difficult to initiate
Easy to use	High UV emission
High quality weld	Major heat losses
Simple current source	
Wire supply separate from welding arc	
Low hydrogen process	

The form of the electrode end is known, and has a major influence on the width of the bead and penetration. When the pressure increases approximately 8 bars, penetration increases, while the width of the bead falls. Penetration is then reduced by modification of the electrode end profile (Ballines). The cathodic task becomes unstable and falls towards the end of the electrode, increasing the power density and causing rapid wear of the electrode.

Finally, when pressure increases, HF arc ignition becomes impossible in a helium atmosphere, even at 2 bars, and the contact ignition technique produces unacceptable tungsten inclusions. The TIG process is therefore limited in a helium or argon atmosphere, remaining a low deposition process. In addition, argon has a major narcotic effect at 150 m and cannot be eliminated by a simple recycling system. It is possible to wear breathing masks, but these are far from practical.

MIG (consumable electrode)

Advantages	Disadvantages
All positions	Special torch
All materials	Recyclable protective gas
No flux	Limited stability range
High rate of deposition	Complex current source
No thickness limit	(Flat characteristic)
High quality weld	Smoke
Self-regulating arc	Projections
Continuous electrode	Turbulent arc
Automation	Difficult wire unwinding adjustment
Good penetration	
Low hydrogen process	

With pressure, the transfer of metal by short circuit develops in the direction of droplet transfer accompanied by a marked increase in smoke and spatter, but major ductility and satisfactory resilience of the fused metal can be obtained when equilibrium between the wire, gas, energy and electrical parameters is satisfactory.

Self-regulation of the length of arc and electrode render this process suitable for automation, but a high level of maintenance is required for the torch. The current source and wire supply may be major difficulties.

Wear of welding bases, spatter and low visibility can make a simple process very difficult to apply, particularly since the equipment is required to exhibit a high level of reliability.

MMA (coated electrode)

Advantages	Disadvantages
Simple current source	Increase in O content
No protective gas	Risk of inclusion
Major preparation tolerances	Presence of slag
All materials	Special electrodes for hyperbaric welding
All positions	Substantial smoke
No limit on thickness	Low visibility
Satisfactory deposition rate	
Quality welds	
Flexibility	
Versatility	

Electrode welding is the least affected by pressure as regards arc and application characteristics. The simplicity, flexibility and versatility with respect to equipment items and consumables makes MMA the most widely used process for hyperbaric welding. Given the manual aspect of the process, the variations in material, thickness, position, are easy to master for welders trained in qualified procedures.

The use of special electrodes gives mechanical characteristics equivalent to those obtained at atmospheric pressure.

There are no particular operational difficulties. From a metallurgical viewpoint, the carbon and hydrogen content of the fused metal requires the surrounding atmosphere to be permanently monitored. The increase in the carbon content as a function of pressure may result in a depth limitation for this method, currently estimated at 300 m.

In conjunction with research laboratories, we have for a number of years been engaged in research programmes on equipment and consumables.

A special basic electrode will enable us to satisfy the mechanical characteristics required under codes API 1104, ES 4515 and Det norsk Veritas procedures, with personnel qualified by Lloyd's Register and DnV.

Current State-of-the-Art

The base metal used was API 5 LX X65. A new electrode has been developed by COMEX for a depth of 300 m. The percentage of carbon in the metal deposited by this electrode increases as much as a normal basic electrode at a depth of 150 m. The percentage residual hydrogen measured in the coating after drying places this electrode in the 'extremely low hydrogen' class. Mechanical tests give extremely high level results, well above the requirements laid down in specifications. Macrographic examination has revealed no defect outside the permitted tolerances.

MIG 300

In parallel to development of the electrode, COMEX have since 1978 developed, refined and obtained approval for a MIG welding procedure for a depth of 300 m. The welding equipment used consists of standard equipment modified for hyperbaric application. The results of the mechanical tests confirm the efficiency of the process, complying with the requirements laid down in specifications.

MIG 400 m and 600 m

The results obtained on investigation of the MIG process for use at 150 m and 400 m demonstrate the advantage of using MIG where a coated electrode cannot yet be used in its place. Another more ambitious study investigates the application of the MIG welding process to depths of 400-600 m. Caisson tests have been carried out to determine the feasibility and quality of MIG welding at 400 and 600 m.

POSSIBLE DEVELOPMENTS

Depth

Techniques now permit operation to a depth of 300 m. The depth of submersion of certain pipes ($>$600 m) is such that they can no longer be repaired by saturation divers. The need for continuity of operation has led to the following potential solutions to the problem of damage.

(1) Duplication of certain runs to ensure redundancy (technologically simple, but costly).
(2) Deep sea repair on damaged sections. COMEX are currently working on the repair solution. Their approach is to break down operation into two phases:

(i) Preparation of the run to be connected or repaired. RMP project (Réparation Mer Profonde or Deep Sea Repair).
(ii) Connection, achieved in one of three ways:
- (a) using welders in welding chambers at atmospheric pressure (WELDAP project);
- (b) mechanical, automated or remote controlled connection;

(c) use of an intelligent welding robot (self-adaptive).

Deep Sea Repair

The object of the Deep Sea Repair project, the fruit of the association between oil companies CFP, SNEA(P) and contractors COMEX and ACB, is to develop equipment and procedures enabling pipe ends to be prepared prior to connection. When pipe damage is localised and the pipe is uncovered, repair can be carried out by the following method:

- roughly cutting the pipe to eliminate stresses;
- setting the lifter and modular preparation system in place (i.e. table, beam, lifting beam, and module);
- clamping the pipe ready for work;
- externally brushing the pipe to remove protection;
- accurately cutting and preparing the chamfer for each end;
- cleaning the inside of the pipe to ensure an effective seal;
- raising the modular system;
- setting the connecting system in place (WELDAP, mechanical connection or welding robot).

Equipment is currently under final construction. Offshore tests (at -300 m) were scheduled to take place in 1984.

THE HYPERBARIC WELDING ROBOT

The field of application is as follows:

- welding on pipes with major geometrical tolerances;
- welding in a gaseous atmosphere at pressures exceeding 20 atmospheres.

Under environmental conditions such as these, only machines capable of adaptation (remote operated and/or with artificial intelligence) could perform such operations. In effect, the preparation of subsea pipes will never attain the quality of onshore alignment and chamfering essential for the use of automatons.

A company such as S.A. COMEX is developing a second generation system capable of adapting automatically to a changing environment. A form recognition sensor operates in the vicinity of the working welding torch.

Appendix 24

Non-destructive Testing Techniques

Testing methods vary in both technique and objectives. They may be classified in various ways. For the purposes of this review, we shall consider these methods according to the degree of technical development, from the simplest, such as dimensional checking by a diver-inspector, to the most complex, such as gamma radiography and its interpretation. A survey of techniques currently being developed will be given at the end of this section.

The majority of the techniques used in subsea inspection are onshore techniques adapted to the hyperbaric environment.

Simple types of inspection include visual inspection and dimensional checking. These are the types of inspection involving the senses, and simple or conventional instruments.

Sophisticated types of inspection include temperature measurement, potential current measurements, magnetoscopy, ultrasonic fault detection, and ionising rays (X or γ). These are types of inspection involving major resources and providing information that requires interpretation. The inspection equipment is used in situ.

It is worth prefacing our examination of remote controlled unit inspection techniques by stating that relatively thorough cleaning is essential, comprising:

- removal of the vegetable and animal cover, i.e. mussels, dogstooth, algae, sponges, etc. which increases the effect of the waves and the weight of the structure, hampers visual inspection, aids corrosion, blocks sea-water intakes, etc.;
- removal of protective coverings, i.e. epoxy, bituminous paint, to reveal the metal surface, to allow non-destructive testing or detailed weld inspection methods to be used.

The main cleaning techniques used for these purposes are high-pressure water-jetting (generally with the injection of calibrated sand as an abrasive), and pneumatic or hydraulic cleaning tools (needle hammers, chisels, rotary brushes, etc.).

HIGH-PRESSURE WATER-JETTING

An approved high-pressure water-jetting system, using a submersible 'Woma Atümat' sand injection system, is provided for preparatory cleaning of areas or parts of underwater structures subjected to detailed visual inspection or non-destructive testing. The jetting hose is fitted with a back-propulsion compensator. The main cleaning orifice and the back-propulsion orifice have interchangeable nozzles to allow different types of jet to be used (a pencil jet, flat diffused jet, etc.) to suit the type of surface or coating.

The jetting hose is also fitted with a flip-flop safety discharge which trips automatically on releasing the trigger. In general, working pressures of up to 667 bars (10 000 psi) are used for cleaning down to bare metal, to allow detailed visual inspection or non-destructive testing.

The sand injection system is normally used in water when cleaning of the protective coating has to be carried out to standard SA 2 1/2-3.

This system comprises a spherical sand reservoir filled with calibrated silica sand, a Woma type 5 nozzle with a special fitting for abrasive sand, and a suitable HP hose. The average cleaning rate is 11 m^2/hr.

TECHNICAL SPECIFICATIONS

Model:	$D150_s$ skid-mounted Hydrojet.
Type:	Positive displacement plunger pump - Volvo Diesel engine and clutch.
Motor:	Volvo Diesel engine TB70B with spark-proof system.
Pump:	Long/Quinn 150 fitted with a stainless steel head for sea-water duty, P 35 mm ceramic plungers.
Rate of flow:	145 l/min (32 gal/min) at 400 bars (6000 psi).
Supply:	900 l integral tank with water intake float. Automatic stop by float at low level.
Checks:	Fuel level etc., automatic trip at high water temperature or drop in oil pressure.
Nozzle:	F.A. Hughes, stainless steel, with retro-jet.
Safety:	Trigger safety. Automatic stop by pressure release on trigger.
Working pressure:	Animal/vegetable cover: 300 bars (45 000 psi). Joint cleaning: 667 bars (10 000 psi).
Overall size:	4256 x 4256 x 1824 m (14' x 14' x 6').
Weight:	5 tonnes.

SIMPLE CHECKS

Visual and Dimensional Checks

These are of course the simplest of the non-destructive tests, using the combination of the marvellous lens in the human eye and intelligence. This is a very widely used method of inspection of structures. Visual inspection of a platform carried out to ascertain the integrity of the primary and secondary structures (i.e. risers, clamps, arrangement of legs, conductor guide frames, etc.) should normally, in the case of routine inspection, comprise the following:

(1) A check on the general configuration of structural members, riser clamps, conductor guides, berths, mudmats, etc.

(2) A general inspection of the platform to ascertain any settlement, buckling, cracking etc., recording lengths, widths, depths and positions. Only the external faces of legs, members, risers and riser supports are inspected in the above investigation. In the event of buckling of a major structural member, the connecting welds of the defective item are inspected to determine the presence or otherwise of perforation, shear or cracking.

(3) General inspection of the platform, particularly in zones of heavy marine cover, to find any pitting corrosion, determining the size, depth and frequency of pitting using a pit gauge, especially at welds and in heat affected zones. The location of pitting is recorded for subsequent ultrasonic examination and electrical potential measurement.

(4) Survey on the state of marine cover classifying according to hard (dogstooth and mussels) or flexible (sponge, sea weeds, etc.). The rate of marine cover is normally established by measuring the thickness of four points at right angles, every 5 m, from floor to surface on each platform pile and riser. An average thickness may also be obtained by measuring the perimeter of the pile or riser in specified locations. This average thickness at the inspection point is then given by the formula:

$$e = 0.5 \left(\frac{P}{\pi} - d\right)$$

where d is the actual diameter of the member inspected.

The level of marine cover may be classified as follows:

- major: full cover thickness 4-10 inches;
- average: full cover thickness 2-4 inches;
- light: full or partial cover less than 2 inches in thickness.

(5) Check on the integrity of the cathode protection system. In general, the position, number and state of fixing of all sacrificial anodes on the platform is reported. The general state of the anode is reported according to the following classification:

- Good: original form retained, with regular reduction in size.
- Average: up to 25% reduction, regular loss retaining the original form.
- Passable: 25-50% reduction and/or irregular form.
- Poor: 50% or more reduction and/or highly irregular form requiring replacement in the near future if possible.

(6) Rubbish survey, identifying and locating all ferrous rubbish less than 2 m from an underwater structure or pipeline. If electrical contact is established, this reduces the effectiveness of the cathodic protection system. All points of contact are carefully inspected to reveal any trace of corrosion, and rubbish is removed or shifted as far away as possible.

(7) Check on the floor profile to determine scouring around and in the vicinity of the piles, legs, conductors, risers and flowlines. A floor profile is also produced and reported on for each face of the platform, measuring the distance from the horizontal bracing to the floor every metre.

The length, width, depth and position of any distortion in the floor potentially evidencing horizontal displacement of the legs is measured. In particular, the state of the riser support bed is reported.

INSTRUMENT OPTICS AND ELECTRONICS

The best extension to visual inspection is a combination of the human eye, instrument optics and electronics, hence photography, macro-photography, black and white or colour television.

Equipment Used

The Nikonos III 35 mm camera is used for all 35 mm underwater photographs in the 0-50 m range, and the Canon F1 in a special housing for deep water photographic work. These cameras are used with Ektachrome 64/200 professional EPR 135-36 films. It is generally recognised that in terms of chromatic fidelity, Kodak E6 transparency films are unrivalled. No other film easily available on the market is recommended for the production of colour slides.

COMEX have developed a system of measuring rods for combined use with a 28 mm F 3.5 lens, to ensure the right distance and depth of field in close-ups. For example, this enables excellent detail to be shown on joint welds. Used in conjunction with the Oceanic Products 2003 electronic flash, this system of photography gives reliable and extremely high quality results.

Technical Specifications

COMEX Nikonos III kit

Camera body. Nikonos III camera, maximum capacity 35 negs. 24 x 36 mm on 35 mm cassette film. Vertically operated horizontal shutter, speed 1/15 to 1/500 seconds and time exposure.
Lenses. (1) 35 mm f 2.5 standard lens for use in air and water. Minimum focal length 0.8 m. Horizontal field angle 39° in water, 53° in air. (2) 28 mm f 3.5 - wide angle for use in water only. Minimum focal length 0.7 m. Horizontal field angle 45° in water.
External close-up. External fitting close-up lens with additional lens. Lens positioning rod, focal rods modified by COMEX for use with standard 80 mm, 35 mm and 28 mm field underwater:

- with 80 mm lens = 53 x 79 mm;
- with 35 mm lens = 109 x 164 mm;
- with 28 mm lens = 144 x 216 mm.

Viewfinders. Frame or optical viewfinder suitable for all above lenses.
Flash. Oceanic Products 2003 with EQ connector sync. NiCd batteries give a minimum of 270 flashes per charge and can be recharged in 4 hours.
Optional extra. Oceanic Products 2002 allowing fully automatic use within certain limits.
Exposure meter. Sekonic L 164 B Marine 11 CdS.
Depth. 50m.

Canon F1

Shutter. Titanium shutter, speed 1 to 1/200 + timed exposure.
Viewfinder. Pentaprism reflex, interchangeable with EE servo viewfinder, amplifier viewfinder T, rapid viewfinder and finder cone.
Information in viewfinder. Exposure meter needle, battery monitoring needle, speed, over/under exposure.
Exposure measurement. In-built CdS photoconductive cell.
Aperture (with ED lens). Automatic diaphragm.
Field. 97% of the true image.
Automation. AE priority shutter speed with viewfinder.
Energy. 1 x 1.3 V mercury battery.
Changeable lenses. FD + FL series.
Film range. 25-3200 ASA.
Flash sync. 1/60 s.
Battery monitoring. In-built.
Dimensions. 99.5 x 146.7 x 49.5.
Weight. Housing : 845 g - with 50 mm lens : 1080 g.

Developing Equipment

A transportable chemicals tank designed by COMEX and fitted with a temperature exchanger is supplied for on-site development of Kodak E6 35 mm transparencies. The exchanger enables maintenance of the temperature critical for optimum results to be monitored. Experience has shown development with the aid of hot water heated baths to give results which are neither regular nor of the best quality available from the Ektachrome professional E6 process.

A light box is provided for examining films after developing. A projection system is not recommended, as this would involve cutting and mounting the exposures individually, with problems of classification and consequently problems in printing contact prints and enlargements.

Technical Specifications

Description. Water tank capable of holding 10 AirVac bottles, contents 2.8 l, for chemicals and rinsing water.
Temperature control. 2 kW heat exchanger coupled to a circulating pump and fitted with a thermostat: 0.1°C.
Dimensions. 130 mm x 350 mm x 420 mm.

Drying Column

Type. Durst UT 100, portable, with drying column.
Description. 2 kW hot/cold filtered air blower in a polythene suspended duct.
Capacity. 3 x 1.70 m lengths - 35 or 70 mm format.

Accessories

Various thermometers, light box, graduated flasks, changing bag, scissors and film storage systems as required.

Daylight Developing Boxes

Paterson developing boxes for 1, 3 or 5 x 35 mm films.

COLOUR TV SYSTEM

The colour television circuit system developed by COMEX is the first application of a high sensitivity professional colour television camera for subsea inspection. The chromatic sensitivity of this system is of primary importance for the assessment of marine cover, corrosion products and the state of the cathodic protection system.

The JVC CY-8800 V camera, which uses a 3-tube dichroic mirror system, improves the resolution of an earlier system by a factor of 2 and light sensitivity by 10.

A high-pressure housing and a surface control system enable the camera to be used to water depths of 300 m.

JVC CY-8800 V - Camera and Housing

The camera (size 465 x 195 mm) is enclosed in a cylindrical housing, dimensions 675 mm in length, 195 mm in diameter. The camera and housing, complete with lighting and gun, have neutral float characteristics in sea-water.

A fully motorised lens system provides for adjustment of the focal length from the surface, and automatic diaphragm aperture. A manual system for automatic aperture correction is also available. The camera is fitted with an automatic system of chromatic adjustment controlled from the surface for comparing variations in spectral distribution.

The camera housing is fitted with a sea-water detector to prevent even the slightest loss in hermetic seal. A series and parallel combination of power regulators stabilises the continuous supply to the camera, irrespective of temperature and length of the umbilical. The camera operates on 13 V d.c. and uses 22 watts. The image is standard 625 lines with the PAL colour system. The resolution is better than 500 lines at the centre of the screen.

Lighting

The camera is normally used with two 500 W halogen quartz lamps mounted at an angle of 120° on the front of the housing. Diverging lenses of varying densities can be used to modify the intensity and spread of the light beam. These lamps are supplied with 120 V a.c., duly insulated and provided with a differential earth circuit breaker.

Surface Control Unit

This unit has been designed and built especially by COMEX for use with the colour camera. The control unit is built into a carrying case and comprises a suitably shock-resistant professional colour 14 in. TV screen, with power pack, control system and diver protection circuits. A high sensitivity differential earth circuit breaker prevents any potentially dangerous insulation breakdown on the lamp supply. The unit operates on 220 V 50/6 Hz, and its power consumption is approximately 1200 W. Its dimensions are 533 x 635 x 482 mm, and its weight is 26 kg.

Electrical Umbilical

A bound composite umbilical is used, comprising a 7 conductor TV cable, De Regt type Ncl 34 for all camera functions, and a dual 2.5 rubber sheathed cable for lighting. The two cables are bound with 6 mm polypropylene cording. Souriau connectors are used for underwater connection, with Ampliversal connectors on the surface.

Technical Specifications (JVC-CY-8800 V)

<u>Lens system</u>. Three RGB system tubes (parallel arrangement with dichroic mirror).
<u>Kine</u>. 2/3 in. electromagnetic focus, type Plumbicon electromagnetic deflection (XQ 1428R, XQ 1428G, XQ 1428B) or Saticon (H9311).
<u>Sync</u>. Internal (SSG manufacture). External (possible with optional unit).
<u>Colour coding</u>. PAL.
<u>DN and colour temperature</u>. Either a DN turret filter or a colour temperature filter (3000K or 6000K).

Appendix 25

Fatigue Strength Check Model for Jacket Type Fixed Platforms

The model descibed below is used for checking the fatigue strength of jacket type fixed platforms in tubular assemblies for deterministic type waves. Fatigue strength is checked according to Miner's law, enabling the cumulative damage D to be calculated, written as:

$$D = \sum_i \frac{ni}{Ni}$$

where: Ni = the number of cycles on failure for a variation in stress of $\Delta\Gamma i$ and ni = the number of cycles undergone by the part for a variation in stress of $\Delta\Gamma i$. The criterion is normally $D \leqslant 1$.

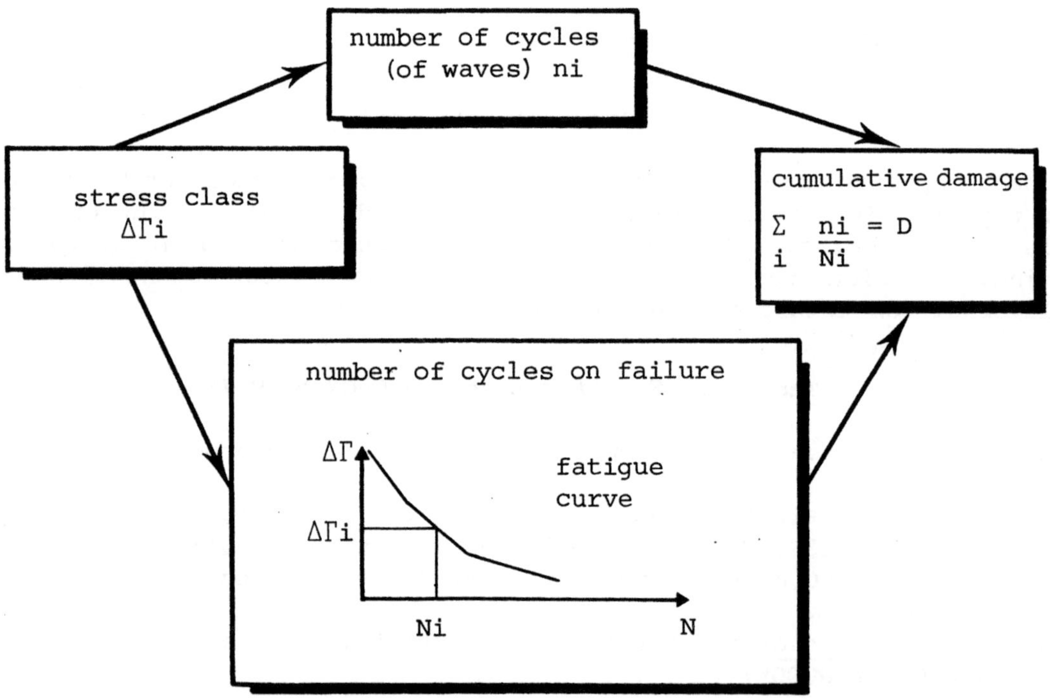

Fig.A.25.1.

Fatigue calculation for a given service life therefore requires a knowledge, for each wave direction, of the characteristics of the various waves (amplitude H and period T) and the corresponding number of cycles n.

As the period of the wave T is considered as a function of the amplitude H, only the amplitude parameter is retained, and knowledge of the law n = f (H) for each direction will suffice.

Establishment of the law giving the variation in stress as a function of amplitude $\Delta\Gamma = f(H)$ is obtained by scanning the waves, and enables the law $\Delta\Gamma = f(n)$ to be determined at a given point.

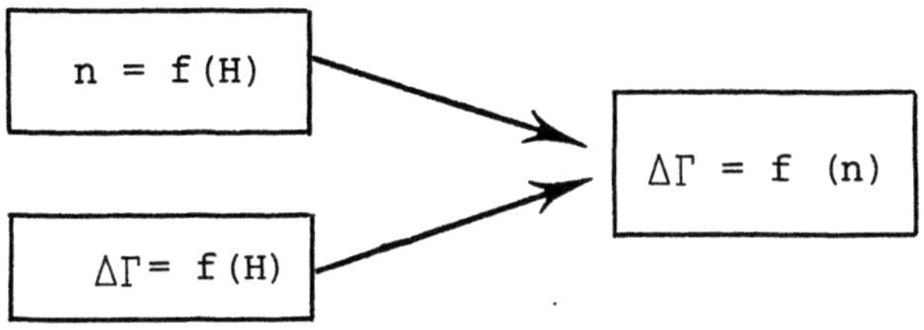

The curve $\Delta\Gamma = f(n)$ is broken down into categories of stress. For each category of stress $\Delta\Gamma i$, the ratio ni/Ni is calculated, where ni is the number of cycles of the wave given by the curve $\Delta\Gamma = f(n)$, and Ni is the ultimate number of cycles given by a fatigue strength curve $\Delta\Gamma = f(Ni)$.

The sum over all stress categories gives the cumulative damage $D = \sum ni/Ni$ over the life in question, at a given point. The stresses at each point are calculated from the cases of wave load analysed, taking into account the stress concentration coefficients and the dynamic amplification coefficients in the case of static analysis.

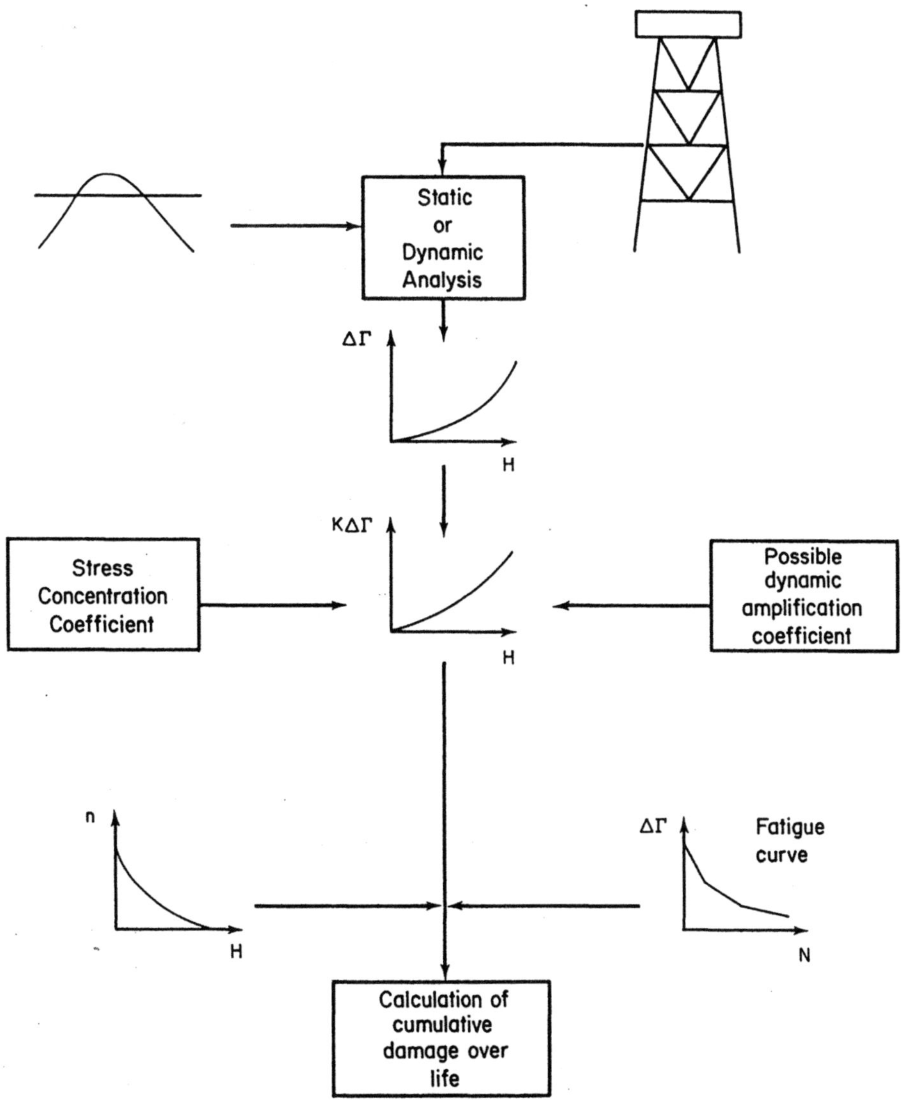

Fig.A.25.2.

The geometry of the structure must first be entered: joint coordinates, incidences and properties of bars. Cases suitable for processing by this programme are as shown in Fig.A.25.3. below.

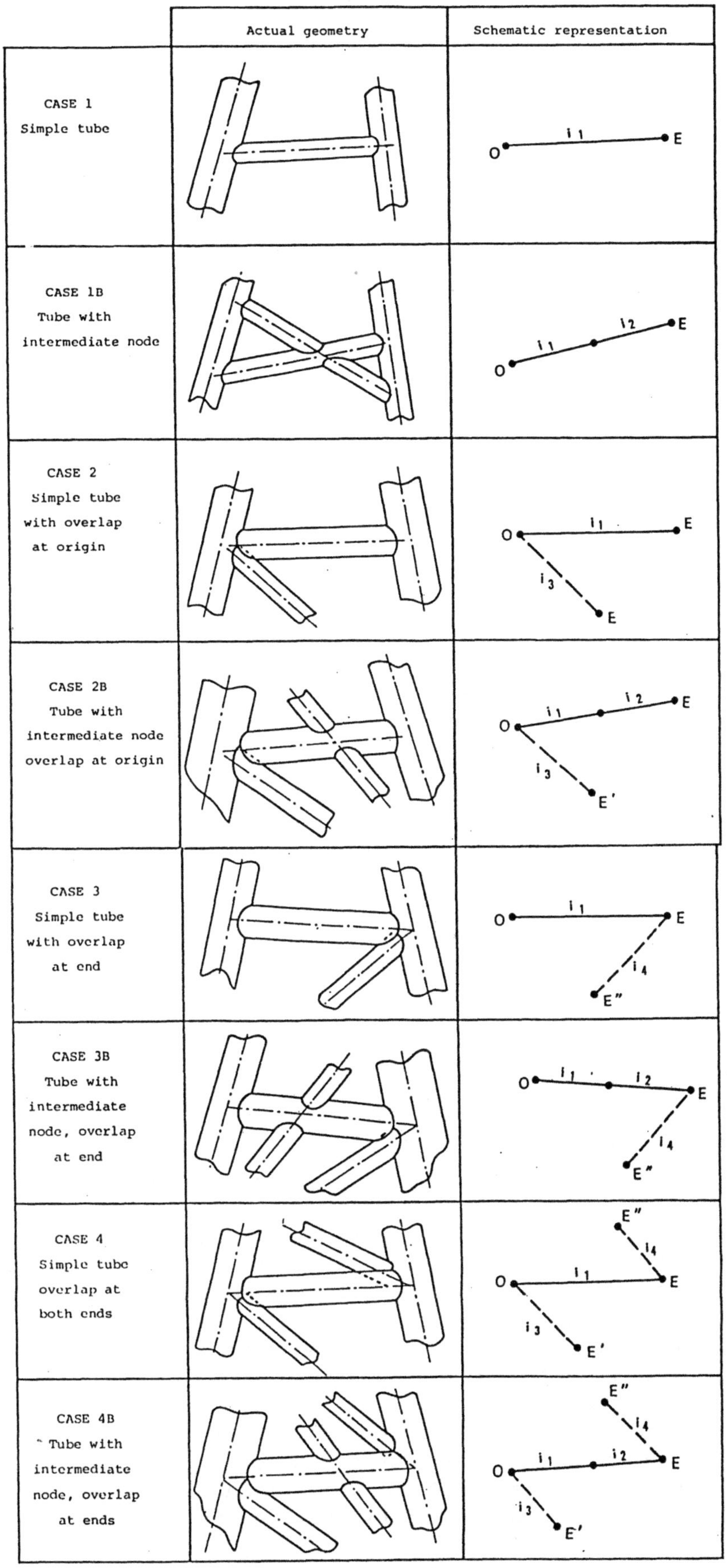
Actual geometry
Schematic representation
CASE 1
Simple tube
CASE 1B
Tube with
intermediate node
CASE 2
Simple tube
with overlap
at origin
CASE 2B
Tube with
intermediate node
overlap at origin
CASE 3
Simple tube
with overlap
at end
CASE 3B
Tube with
intermediate
node, overlap
at end
CASE 4
Simple tube
overlap at
both ends
CASE 4B
Tube with
intermediate
node, overlap
at ends

Fig.A.25.3.

The program will require only the number of the tube to be cut i_1 or $i_1 + i_2$. The program will then:

(a) search for receiving members at node origin 0, then at end node E;

(b) calculate the parameters for lever adjustment (A, B, C) for intersections;

- with receiving members;
- with any overlapped members i_3 and/or i_4 which must also be input into the program.

General Form of Command

SEISMS (Description)

(Description)

-
-
-
-
-

(Description)

Where

Description = list (ØR list .... (ØR list))

$$\text{list} = \left[\begin{array}{llllll} * & 'a_1' & (v_1) & & & \\ & n_i & \left[\begin{array}{c} \underline{\text{TØ}} \\ \underline{\text{TO}} \\ \underline{\text{TH}} \\ \underline{\text{A}} \end{array} \right] & n_2 & \left(\left[\begin{array}{c} \underline{\text{PAR}} \\ \underline{\text{PAS}} \\ \underline{\text{BY}} \\ \underline{\text{STEP}} \end{array} \right] \quad n_3 \right) & (v_1) \\ & i & (v_1) & & & \end{array} \right]$$

Components

$'a_1'$ = alphanumerical identifier of loadings or groups.
n_1, n_2, i = numerical identifiers of loadings or groups.
n_3 = optional whole number (standard value 1).
v_1 = optional real number (standard value 1).

Explanations

(1) This command allows SEISMIC assumptions to be classified.

(2) Each list, contained between two 'ORs', comprises cases which are compatible with one another and will consequently be systematically taken into account in the cumulative total.

(3) The lists are incompatible with one another.

(4) The components of each list may be:

- loadings previously defined by commands LØADING or LØADING CØMBINATION. These loadings must have been active over the analysis for which the results are to be combined. In addition, they must be included in the last LØADING LIST command used.
- groups (see this command) previously defined.

These loadings (or groups of loadings) may possibly be multiplied by a coefficient v_1 enabling the non-weighted actual cases to be defined (e.g. from unit cases): v_1 has nothing to do with the statutory weighting coefficients.

(5) When seeking statutory combinations, the weighting applied to the permanent loadings thus defined are as follows:

- 4/3, 17/12 or 3/2 for combinations with normal climatic loads.
- 1 for combinations with extreme climatic loads.

(6) This command is optional.

(7) In ADDITION and DELETION modes, if the command is repeated, the new data are added to the existing data.

(8) In CHANGE mode, the new data replace the existing data.

(9) Command 'SEISMES' must be preceded by command STIFFNESS and possibly command STRESS ('SOLLICITATIONS') if this is used.

Notes

(1) The notes relating to command DECK ('PONT') are equally valid for command 'SEISMES'.

(2) The current version of command 'SEISMES' is provisional. The weightings adopted are subject to later modification.

Appendix 26

Safety of Treatment Installations*

INSTALLATION LAYOUT

General

The problem of layout of installations for fire safety purposes is related to the following three principles:

(1) all equipment items exhibiting equivalent risk levels to be grouped within the same geographical zones;
(2) physical separation to be provided between geographical zones thus defined, to avoid propagation of any incident from a high risk zone to a lower risk zone;
(3) possible subdivision to be provided for geographical zones thus defined, to limit the consequences of the maximum allowable incident (assumed limited to one zone or one subdivision).

Application of these principles to onshore installations does not generally pose problems, since the space available is sufficient to comply with the necessary safe distances. This does not apply to offshore platforms.

Considerations on the General Organisation of the Platform

General Considerations

Offshore, the space allowed for oil installations is always reduced to the strict minimum and the distances between equipment items are limited to those strictly necessary for maintenance operations. In addition, offshore environmental conditions are such that it is always necessary to have semi-closed type installations.

Indeed, when an open platform is referred to, this is in fact a platform which has only side openings to provide a certain natural ventilation for the modules.

Under these conditions, the imposed solution is to group equipment items of the same type and exhibiting the same hazards within the same areas, and to arrange the relevant areas in such a way that gas sources and sources of ignition are as far apart as possible, while at the same time ensuring that partitioning and ventilation are not over-complicated.

API RP2G provides for six families of plant or installation corresponding to size geographical zones, broken down into three groups.

1. (a) the well zone;
   (b) the hydrocarbon production and treatment zone, excluding combustion machinery;
   (c) the oil storage zones.

---

* Extract from a study by AFTP.

These three zones are hazardous and constitute the first group. Potential sources of ignition must be limited.

2. (a) The combustion machinery zone;
   (b) the machinery zone.

In these zones, fuel has to be limited (fire). They constitute the second group.

3. (a) living quarters and emergency equipment.

This zone constitutes the third group.

Naturally, fire and explosion prevention is not the only concern of the engineer responsible for platform layout (weight distribution, access problems, maintenance, pollution control). It is, however, sufficiently important for the following rules to be taken into consideration.

Vertical Arrangement

Vertically, the lower level (cellar deck), which can be open to the bottom and on four sides, should preferably be used for installations comprising first degree sources not requiring permanent attendance by personnel, and not particularly sensitive to sea spray (e.g., storage tanks and sump tanks, and pig traps).

Pig traps are to be located with doors outwards (particularly receiving traps). This arrangement is designed to minimise the consequences of door ejection.

Particular attention is to be paid to the position of the exhausts from thermal motors or high-powered turbines. These exhausts generate atmospheric disturbance and extremely dangerous variations in buoyancy in the vicinity of heliports.

These exhausts are to be located outside the helicopter approach zone and never on the prevailing windward side of the helipad.

Horizontal Arrangement

The horizontal arrangement of the platform is based on the 'Neapolitan slice' principle. Any hazardous zone must be located downwind of a less hazardous zone under prevailing wind conditions. Each zone is thus located in order of increasing hazard classification, in the direction of the prevailing wind. The well zone, a first degree source, must be located opposite the living quarters, and the entire assembly must be orientated in relation to the prevailing winds in such a way that the living quarters are under no circumstances situated to leeward of the well zone under prevailing wind direction conditions.

To provide good natural ventilation, the wellhead zone must be largely open to the bottom and on at least two sides. This lateral positioning of the openings also allows access by fire boats.

Starting from the wellhead zone, the layout sequence for a production platform with living quarters could be:

- the production installations section (separation, de-sanding, gas treatment and recompression, crude pumping) - these are not in principle first degree sources (zone);
- the motive section (whether or not combustion type);
- the section containing living quarters, electrical installations, emergency generators, fire pump, etc.

Each of these sections must be separated from the others by a fire wall (required to be gas-tight), type A60 minimum under the SOLAS 74 convention.

The location of the heliport, situated in the lowest risk zone, is to be determined as a function of prevailing winds and the free sector for flight approach.

This is an extremely simplified outline, and divisions are of course required within each section (e.g., to provide safe zones for a control room or motorised fire pump - a minimum of two are required on board the platform, to be as far apart as possible).

It should also be borne in mind that the living quarters themselves represent a source of danger. Equipment such as control rooms or fire pumps or emergency electricity sources cannot be installed without taking precautions. These items are to be located in A60 type compartments under the SOLAS 74 convention.

The application of these various principles leads to a typical layout for a drilling and production platform as described in Figs.A.26.1. and A.26.2.

Methodology

General

The ideal safety solution consists of grouping together all items of an equivalent risk level on a separate platform. This would give a set of platforms, each having a particular risk level, and interconnected by walkways. The safe distances required between platforms of differing risks can easily be obtained.

This solution is not realistic, in view of the installation costs, and we are generally compelled to adopt intermediate solutions, depending on the depth of water at the place of installation:

- a single platform for major water depths (North Sea type);
- multiple platforms for shallow water depths (Persian Gulf type).

Based on cost requirements, the designer must determine the optimum number of supports for each conceptual design. Safety requirements must then be taken into account enabling the respective positions of the platforms and the types of equipment or installation carried to be decided. To assist in this procedure, a certain number of criteria may be defined. It is not our intention here to provide an exhaustive list of safety criteria, but instead to adopt those which seem to us to be common to the majority of offshore oil installations.

Safety Criteria

Principle A. Petroleum equipment and installations may be broken down into three distinct parts.

(1) Drilling/production, mud treatment and/or wellhead zone, represented by the abbreviation 'WH'. This zone is considered to constitute an obstacle to aircraft (drilling mast).
(2) The liquid and gaseous hydrocarbon treatment and production zone, represented by the appreviation P. This zone may comprise obstacles such as cranes or turbine stacks.
(3) Utilities and living quarters zone represented by the abbreviation A. This generally has the heliport on the top.

For the purposes of this document, we shall use the term 'platform' to signify the three functional parts described above, without prejudging their position on separate or common structures.

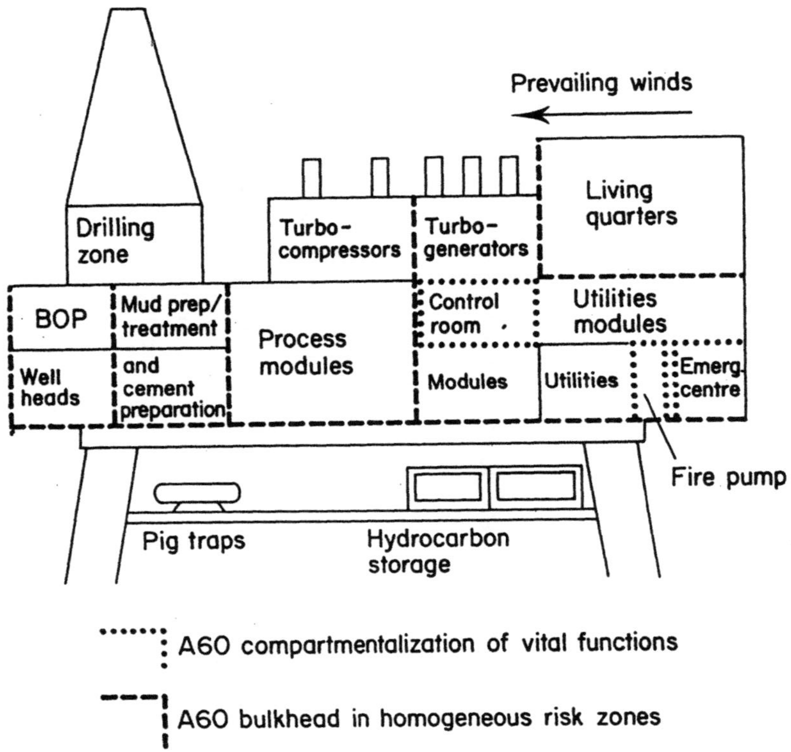

Fig.A.26.1.

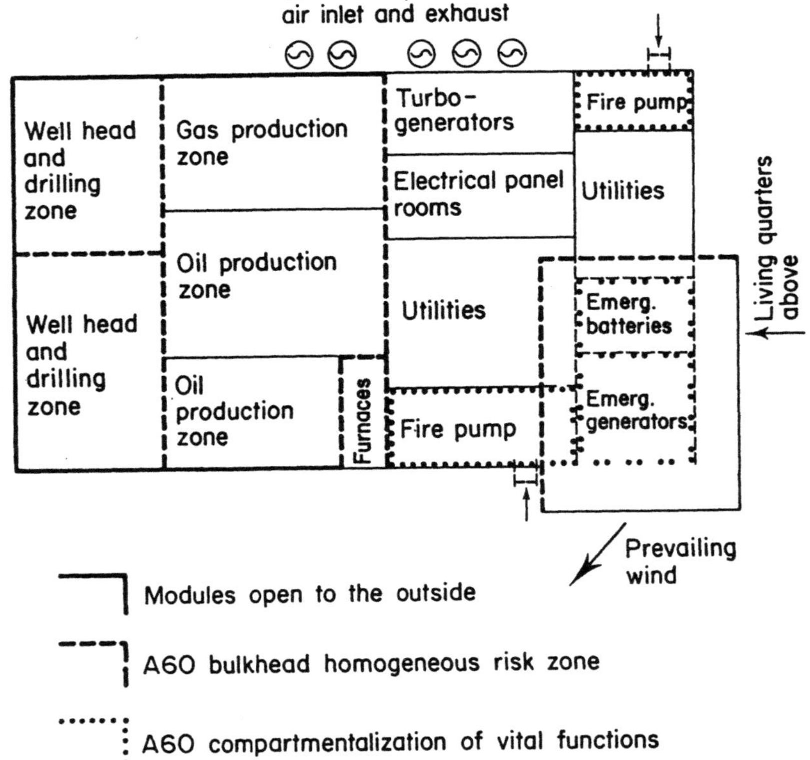

Fig.A.26.2.

Principle B. The platform environment is described by the following elements:

- prevailing wind direction;
- current direction (if any).

A permanent or other flare is shown, symbolising the most probable potential point of ignition. It constitutes an obstacle to aircraft and shipping in view of its structure and the thermal radiation liable to be generated.

Principle C. The living quarters serve as a refuge in the event of an uncontrollable incident in the wellhead and/or production/treatment zone. Personnel retreat to this zone to await assistance and evacuation.

Principle D. Each proposed layout is subject to evaluation of the seven safety criteria specified below. For each criterion, an assessment is given according to the following principle:

- + 1 favourable position taking into account environmental constraints;
- 0 acceptable position taking into account environmental constraints;
- - 1 unacceptable position taking into account environental constraints.

The best arrangement is not that which gives the highest algebraic sum, but that containing no negative terms.

Propagation criterion. Propagation is the passage of fire from the zone of outbreak to an adjacent zone under the effect of thermal radiation of the flames, spread by the wind, shock waves, explosion waves, etc.

Applying the first principle criterion 'any hazardous zone must be situated downwind of a lower hazard zone, in the prevailing wind direction' (see Figs.A.26.3. and 4). However, the position of platforms in a transverse direction in relation to the prevailing wind is regarded as acceptable.

Ignition criteria. The assumed source of ignition in the event of major hydrocarbon leak is the flare. We shall consider two particular criteria. One takes into account the risk of a liquid hydrocarbon slick on the sea, driven by the wind. The wind is in fact regarded as the predominant driving factor for hydrocarbon slicks.

SAFE DISTANCES BETWEEN INSTALLATIONS

General

In this section, we shall briefly describe the rules to be observed in determining the isolation distances to be maintained between two installations or between installations and third parties where at least one could be the originator of a major hydrocarbon leak. The term 'installation' is used here to signify plant and equipment classified into six families under API RP 2G. These families will henceforth be termed installations of the type: T.-1/2/3/4/5A/5B/6 in accordance with the said table.

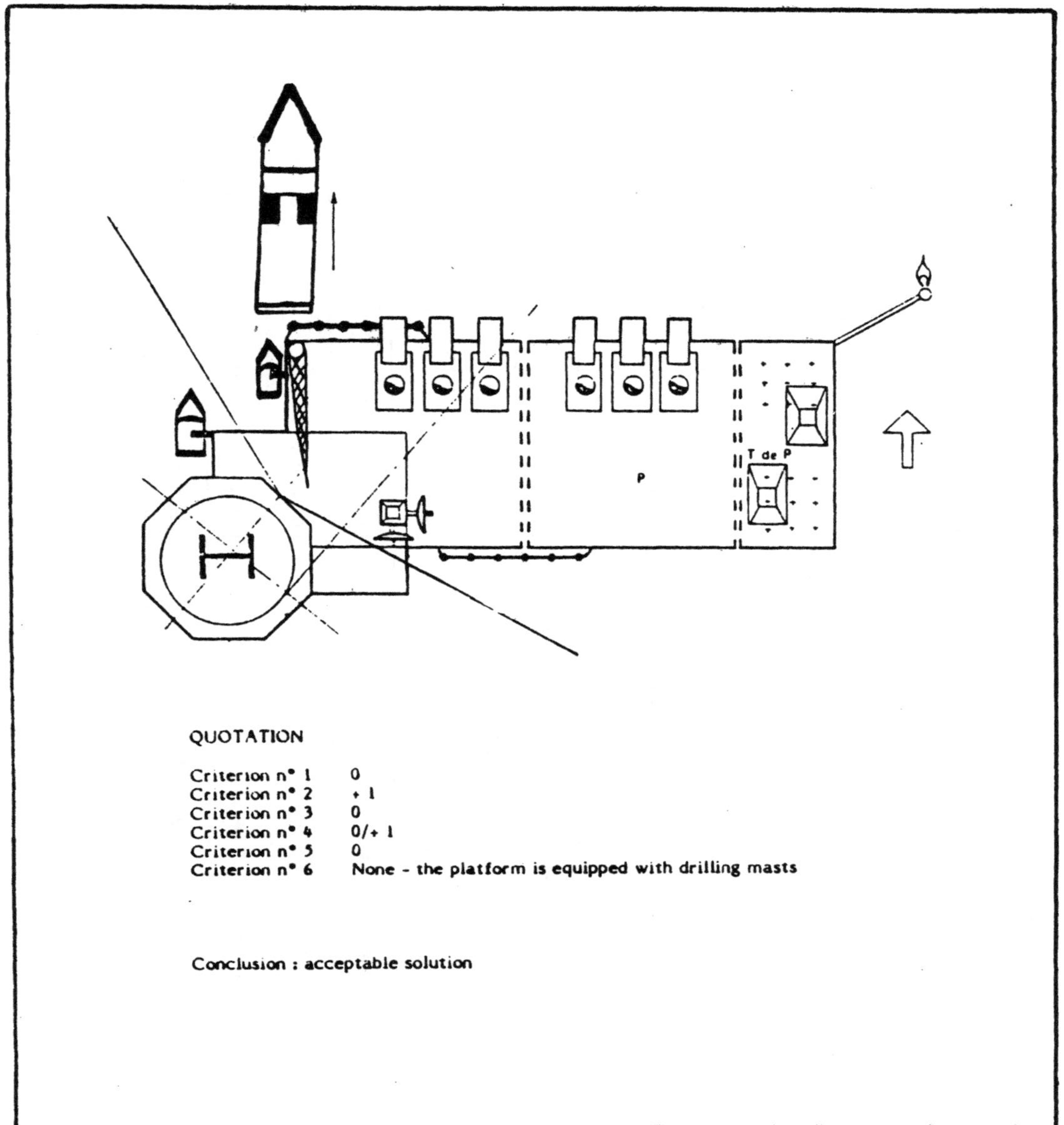

Fig.A.26.3. Examples of application
- single platform in deep water

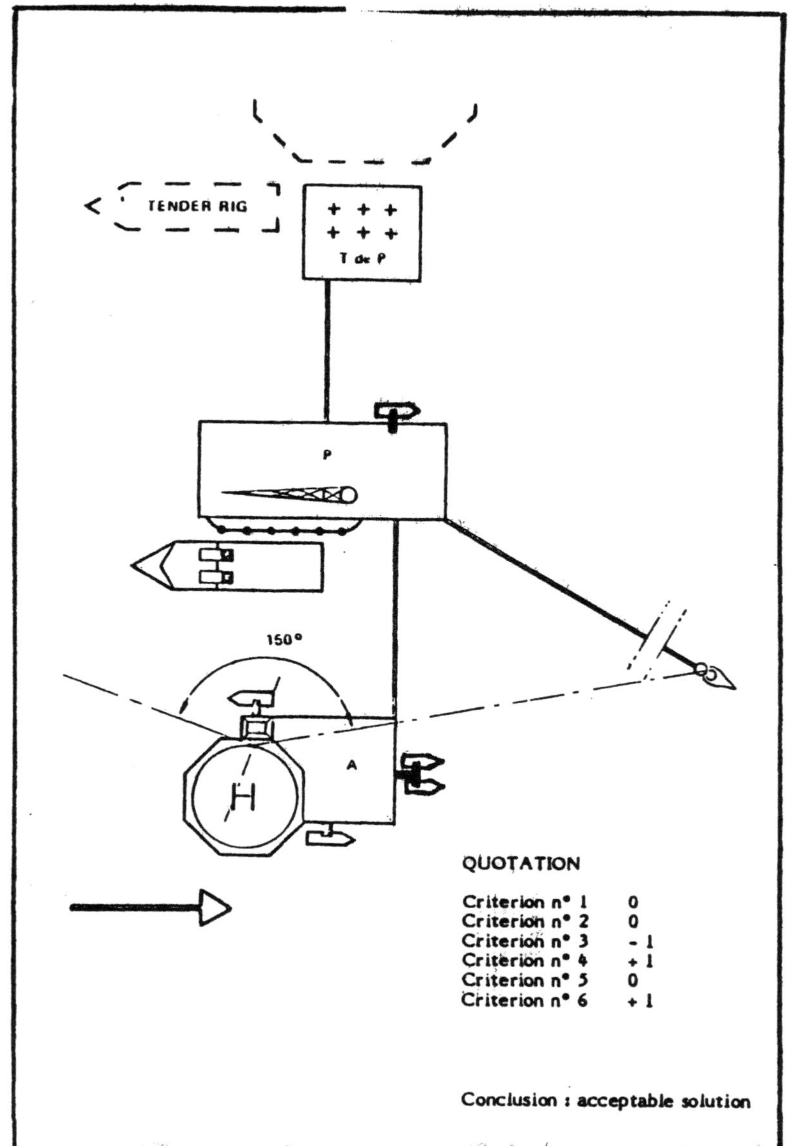

Fig.A.26.4. Examples of application
- multiple platforms in shallow water

Spacing Determination Criteria

Required distances between installations vary according to:

- the potential effects of hazards liable to occur during a major leak;
- the type of operation performed on the hydrocarbons;
- the efficiency of screens inserted between the installations.

The potential effects of hazards regarded as characteristic are as follows:

- thermal flux caused by an ignited hydrocarbon leak - expressed in kilowatts/m^2;
- maximum excess pressure exerted by explosion of a cloud of gaseous hydrocarbon and air, expressed in bars.

The required spacing or screens are considered as a function of the risk, i.e.:

- the potential effects of characteristic hazards, i.e. the gravity of the effects of a major leak;
- the probability of such accident.

Hazard Zones

Hazard Zone Classification

Definitions. For any item of petroleum equipment potentially at the centre of a hydrocarbon leak, the hydrocarbon contained is the determining factor in the hazard zone.

Where valve closure is not instantaneous, the rate of flow during the closing operation is taken into account.

The notion of hazard zone does not apply in units exhibiting the same level of risk: these units are stated as compatible in Table A.26.1.

Table A.26.1. Compatibility of equipment in units exhibiting the same risk level.

Installation group	Type 1	Type 2	Type 3	Type 4
Type 1	Compatible	Compatible	Incompatible	Incompatible
Type 2	Compatible	Compatible	Compatible	Incompatible
Type 3	Incompatible	Compatible	Compatible	Incompatible
Type 4	Incompatible	Incompatible	Incompatible	Compatible
Type 5A	Incompatible	Compatible	Incompatible	Incompatible
Type 5B	Incompatible	Incompatible	Incompatible	Compatible
Type 6	Incompatible	Incompatible	Incompatible	Incompatible

Installation group	Type 5A	Type 5B	Type 6
Type 1	Incompatible	Incompatible	Incompatible
Type 2	Compatible	Incompatible	Incompatible
Type 3	Incompatible	Incompatible	Incompatible
Type 4	Incompatible	Compatible	Incompatible
Type 5A	Compatible	Incompatible	Incompatible
Type .5B	Incompatible	Compatible	Compatible
Type 6	Incompatible	Compatible	Compatible

Gravity category. Hazard zones are further classified into five gravity zones according to the probable gravity of hazard for persons and property. These categories are noted as Z1/Z2/Z3/Z4/Z5 in decreasing order of gravity, and described in Table A.26.2.

Table A.26.2. Gravity categories.

Zone designation	Z1	Z2	Z3	Z4	Z5
Foreseeble injury to persons	Fatal injury in less than 50% of cases	Serious possibly fatal injuries	Injuries	Possi-bilities of injuries	Very slight possibility of minor injuries
Foreseeable damage to property	Very serious damage	Major damage	Moderate and minor damage	Minor damage	Very minor damage

Extent of hazard zones. The extent of hazard zones is essentially governed by the configuration of the zone and equipment, the means of protection provided and the nature of danger covered, i.e. fire or explosion.

Thus the safe distances corresponding to each gravity category is calculated from the effects of fires and explosions.

Fire and explosion cases are investigated independently for each installation liable to cause a major leak.

The extent of hazard zones is determined from Table A.26.3.

Table A.26.3. Determination of hazard zones.

Hazard zone	**Fire** **Ø - Thermal flux released** **kilowatt/m²**	**Explosion** **P = max. surge** **bar**
Z1	Ø >15	P>0.6
Z2	15 >Ø >9.6	0.6>P>0.3
Z3	9.6 >Ø >4.8	0.3>P>0.1
Z4	4.8 >Ø >1.4	0.1>P>0.05
Z5	Ø <1.4	P<0.05

Production of hazard zone drawings. Having calculated the hazard zone radii with the aid of Table A.26.4., a hazard zone drawing must be prepared from the proposed layout.

Table A.26.4. Probability classes

Probability class	**P1**	**P2**	**P3**	**P4**	**P5**
Definition	Extremely rare	Very rare	Rare	Quite frequent	Frequent
Probability of occurrence per year T	$T<10^{-4}$	$10^{-4}<T<10^{-1}$	$10^{-3}<T<10^{-1}$	$10^{-2}<T<10^{-1}$	$T>10^{-1}$

Hazard zone radii are to be drawn from possible outside volumes specific to each risk.

(a) Fire risk is broken down into two categories:

- Fire of the hydrocarbon slick in the liquid phase. The radius of the hazard zone is determined by the perimeter of the maximum area liable to be affected by the spillage.
- Gas fire. The source will be assumed to be localised, and hazard zone radii will be determined by the outside flame centre perimeter.

(b) For explosion risk, the following method is to be used:

- calculation of the maximum extent of the explosive cloud;
- measurement of the hazard radii from the outside area of the cloud.

Probability of Major Leak

Each installation may be assigned a probability of occurrence of a major leak. For each family of installations (defined in Table A.26.1.), it is possible to define a probability of occurrence of incidents on condition that the installation does not incorporate any innovation or unusual design or construction features. An example is given in Table A.26.5.

Table A.26.5. Estimate of failure rates of particular families of installations.

Installation family	**Main installations**	**Probability class**
Type 1	Wellheads - drilling/workover/wireline Wellheads - instrumentation	P4
	Wellheads - production Other type 1 installations	P3

This probability must however be evaluated with a greater degree of precision using risk analysis methods, in the following cases:

(1) Wherever use of the table causes the feasibility of the project to be questioned.
(2) For all new equipment items. The definition of probability classes is quantified as the probability of occurrence per year, thus allowing the result obtained by risk analysis to be incorporated in this methodology.

Maximum Risks Allowable in Hazard Zones

For the purposes of safeguarding individuals and property, it is possible to determine allowable layouts for the various types of installations within geographical zones, identified by the hazard zone in which they are located, and the probability of occurrence of such hazard.

Installation Layout Rules

Figure A.26.5. gives a possible layout for the various categories of installation in a zone with a hazard Z3P3 where:

(1) index 3 of Z3 indicates the gravity of the hazards involved;
(2) degree 3 of P3 indicates the probability of major leakage giving rise to such hazard.

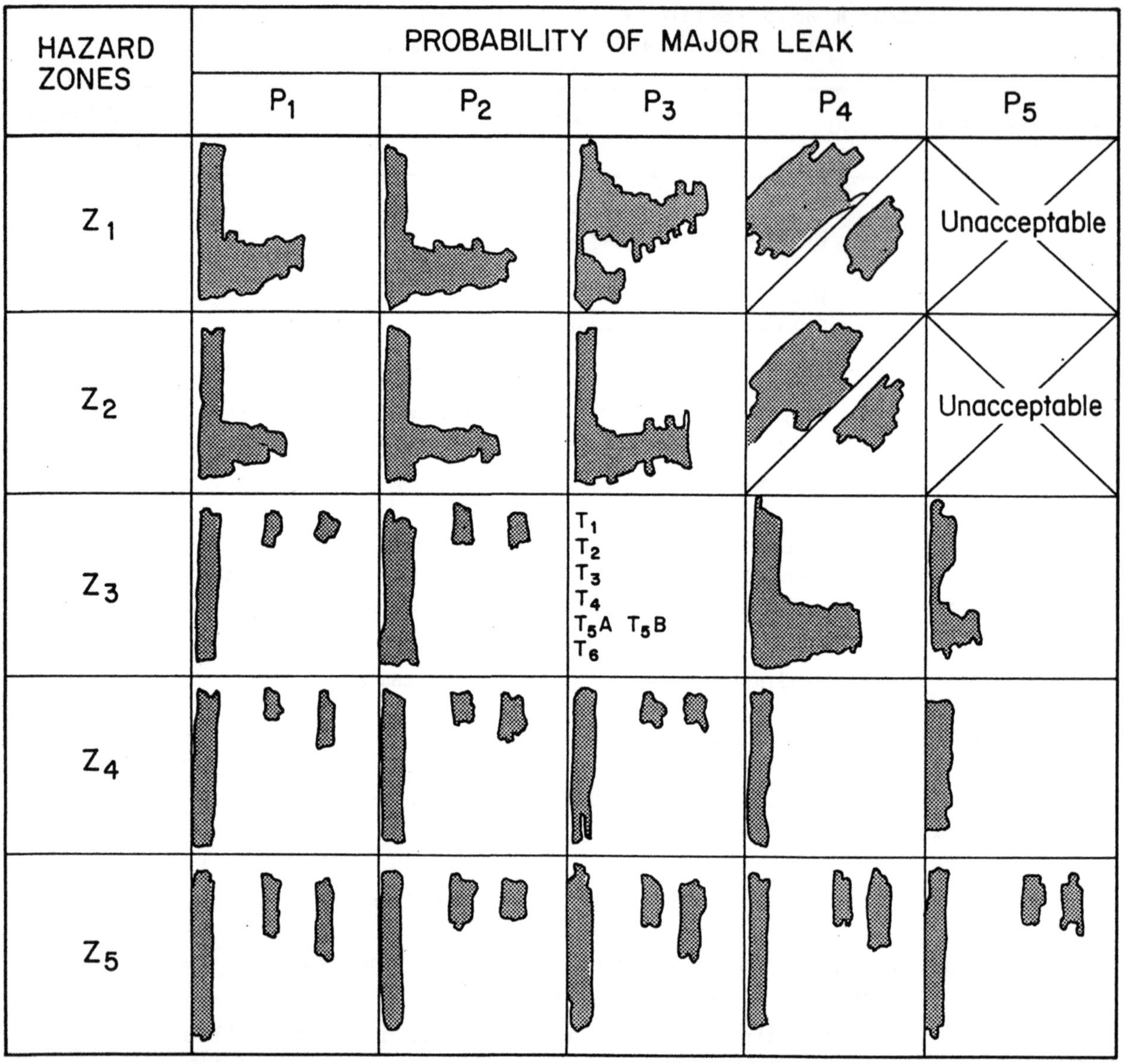
HAZARD ZONES
PROBABILITY OF MAJOR LEAK
P1
P2
P3
P4
P5
Z1
Unacceptable
Z2
Unacceptable
T1
T2
T3
T4
T5A T5B
T6
Z3
Z4
Z5

Fig.A.26.5.

GPSR Compliance

The European Union's (EU) General Product Safety Regulation (GPSR) is a set of rules that requires consumer products to be safe and our obligations to ensure this.

If you have any concerns about our products, you can contact us on ProductSafety@springernature.com

In case Publisher is established outside the EU, the EU authorized representative is:

Springer Nature Customer Service Center GmbH
Europaplatz 3
69115 Heidelberg, Germany

Batch number: 10585057

Printed by Printforce, the Netherlands